T0137033

High Accuracy Surface Modeling Method: The Robustness

Na Zhao · TianXiang Yue

High Accuracy Surface Modeling Method: The Robustness

 Springer

Na Zhao
State Key Laboratory of Resources
and Environmental Information System
Institute of Geographic Sciences
and Natural Resources Research
Chinese Academy of Sciences
Beijing, China

TianXiang Yue
State Key Laboratory of Resources
and Environmental Information System
Institute of Geographic Sciences
and Natural Resources Research
Chinese Academy of Sciences
Beijing, China

ISBN 978-981-16-4029-2   ISBN 978-981-16-4027-8   (eBook)
https://doi.org/10.1007/978-981-16-4027-8

This Springer imprint is published by the registered company Springer Nature Singapore Pte Ltd.
The registered company address is: 152 Beach Road, #21-01/04 Gateway East, Singapore 189721, Singapore

Preface

Based on the principal theorem of surface theory, high-accuracy surface modeling (HASM) transforms the problem of surface simulation into algebraic equations through constraint control of sampling points. Existing studies show that HASM can eliminate the errors that have caused problems for geographic information systems (GISs) for a long time, and its accuracy is higher than that of classical interpolation methods. However, HASM still has several theoretical limitations. For instance, HASM currently relies on other methods to generate the driving field and is sensitive to the selection of driving field, while the simulation accuracy with an arbitrarily selected driving field is low. Moreover, the stopping criteria for iteration in HASM have no theoretical basis, and the computational speed of HASM is slow, with an oscillating boundary. These problems severely hinder the application of HASM. To improve the performance of HASM and enhance its robustness and stability, the following research work based on the principal theorem of surface theory was carried out.

Through the introduction of an equation satisfied by the mixed partial derivative in the system of partial differential equations of the surface and the use of a stable difference scheme for the mixed partial derivative, a more accurate HASM method is proposed, the modern HASM method (HASM.MOD). The influence of a high-order finite difference scheme on the simulation accuracy of HASM is studied. Through a comparison of the influences of the mixed partial derivative term and high-order finite difference scheme on the simulation results, the reasons for the improvement in the accuracy of HASM are explored. Based on the principal theorem of surface theory and the introduction of the Gauss–Codazzi equation into HASM, quantitative indexes of the stopping criteria for iteration in HASM are proposed, which avoids the arbitrary selection of stopping criteria for iteration and promotes the intelligent development of HASM. In the improved HASM.MOD, the boundaries are processed in the same way as the interior of the simulation region, the boundaries are no longer considered separately, and the high-order difference discrete scheme is adopted for the simulation region. A disadvantage of traditional HASM is its dependence on surface interpolation methods, so the sensitivity of traditional HASM and modern HASM to the selection of the driving field is studied, the properties of the system matrix of traditional HASM and HASM.MOD equations are analyzed, and the reasons why HASM.MOD is insensitive to the selection of the driving field are explored. In addition, the simulation results of HASM.MOD with an arbitrary driving field, especially a zero driving field, are verified through numerical experiments and case studies. According to the real application conditions, the influence of sampling information on the simulation results of HASM.MOD is studied from two perspectives, i.e., sampling density and sampling error. Last, based on the characteristics of the coefficient matrix of the HASM.MOD equations and the conjugate gradient (CG) method, the original equations are transformed into well-conditioned equivalent linear equations by selecting the optimal preconditioning operator so that a rapid method for solving the HASM.MOD equation is given, and the parallel programming of HASM.MOD is realized with a message passing interface (MPI).

In summary, based on the principal theorem of surface theory, this book presents HASM.MOD, a more accurate modern HASM method, to improve the robustness and performance of HASM. With temperature, precipitation, and mathematical surface as the experimental objects, the performance of HASM.MOD is verified. The improvement in the simulation accuracy of HASM can be mainly attributed to the introduction of the third equation (the equation satisfied by the mixed partial derivative) in the Gauss equations, as well as the high-accuracy difference scheme. The simulation errors at the boundaries are reduced, and the oscillating boundary of HASM is eliminated. With high robustness, HASM.MOD is a surface simulation method with high accuracy and is completely independent of the interpolation method, and the sensitivity of HASM.MOD to the selection of driving field is eliminated; thereby, the practicability of HASM.MOD is improved. The influence of the sampling information on the simulation results of HASM.MOD shows that as the sampling percentage increases, the simulation accuracy of HASM.MOD gradually increases. As the number of sampling points increases, the representativeness of the sampling points should be considered, and the simulation accuracy of HASM.MOD declines significantly with increasing sampling error. The HASM.MOD can be solved rapidly based on the CG method and the parallel algorithm based on an MPI, which improves the performance of HASM. Finally, the improved HASM is applied to temperature, precipitation, sunshine percentage, and evaporation.

Beijing, China   Na Zhao
                                                                                                          TianXiang Yue

Acknowledgements This book is supported by the National Natural Science Foundation of China and the Strategic Priority Research Program (A) of the Chinese Academy of Sciences (Grant No. 41930647, 42071374, and XDA20030203).

Contents

1 Introduction 1
  1.1   Research Background and Significance 1
      1.1.1   Surface Modelling 1
      1.1.2   HASM of Climatic Elements 2
  1.2   Research Background 3
      1.2.1   Common Surface Reconstruction Methods 3
      1.2.2   High Accuracy Surface Modelling (HASM) 9
      1.2.3   Main Spatial Interpolation Methods of Climatic
              Elements 20
  1.3   Research Objectives, Contents, and Methods 28
      1.3.1   Research Objectives 28
      1.3.2   Research Contents and Methods 29
  References 30

2 Modern HASM Method 41
  2.1   Improvements in Traditional HASM Methods 41
      2.1.1   Numerical Simulation 47
      2.1.2   Case Study 55
  2.2   Reasons for Accuracy Improvement 62
  2.3   Boundary Value Problem 63
  2.4   Stopping Criteria for Iteration in HASM 65
  2.5   Summary 68
  References 69

3 Sensitivity of HASM to the Selection of the Driving Field 71
  3.1   Sensitivity of Traditional HASM to the Selection
      of the Driving Field 71
  3.2   Comparison of the Sensitivity of HASM and HASM.MOD
      to the Selection of the Driving Field 74
      3.2.1   Numerical Simulation 74
      3.2.2   Case Study 75
  3.3   Summary 82

**4 Influence of Sampling Information on the Performance
    of HASM.MOD** 85
  4.1 Influence of the Sampling Ratio on Simulation Accuracy
    of HASM.MOD 85
    4.1.1 Numerical Simulation 86
    4.1.2 Case Study 88
  4.2 Influence of the Sampling Error on the Simulation Accuracy
    of HASM.MOD 97
    4.2.1 Numerical Simulation 97
    4.2.2 Case Study 100
  4.3 Summary 107
  References 109

5 Fast Computation and Parallel Computing of HASM.MOD 111
  5.1 Fast Computation Method for HASM.MOD 111
    5.1.1 Incomplete Cholesky Conjugate Gradient (ICCG) 114
    5.1.2 Ssymmetric Successive Over
    Relaxation-Preconditioned Conjugate Gradient
    (SSORCG) 115
    5.1.3 Numerical Simulation 116
  5.2 Parallelization of HASM.MOD 118
  5.3 Summary 122
  References 123

**6 High Accuracy Surface Modelling of Average Seasonal
    Precipitation in China Over a Recent Period of 60 Years** 125
  6.1 Introduction 125
  6.2 Data 127
  6.3 Methods 128
    6.3.1 Polynomial Regression 128
    6.3.2 HASM 131
  6.4 Model Verification and Result Analysis 132
    6.4.1 Model Verification 132
    6.4.2 Results 134
  6.5 Results and Discussion 137
  References 138

7 HASM of the Percentage of Sunshine in China 139
  7.1 Introduction 139
  7.2 Data and Study Area 140
  7.3 Simulation of Average Monthly Percentage of Sunshine 141
  7.4 Conclusion 144
  References 144

8 Simulation of Potential Evapotranspiration in the Heihe River Basin by HASM 147
  8.1 Introduction 147
  8.2 Study Area and Data Source 148
  8.3 Methods 150
    8.3.1 Polynomial Regression and Stepwise Regression 150
    8.3.2 HASM 151
  8.4 Results 159
    8.4.1 Accuracy 159
    8.4.2 Comparison of the Potential ET in the Heihe River Basin Simulated by HASM and Kriging 163
  8.5 Results and Discussion 164
  References 166

9 HASM-Based Downscaling Simulation of Temperature and Precipitation and Scenario Prediction in the Heihe River Basin 169
  9.1 Introduction 170
  9.2 Data and Methods 171
    9.2.1 Data 171
    9.2.2 Methods 171
  9.3 Comparison of the CMIP5 Baseline Data and Meteorological Observation Data 175
  9.4 Downscaling Simulation of Future Temperature and Precipitation in the Heihe River Basin Under Different Scenarios 179
  9.5 Conclusion 186
  References 186

List of Figures

Fig. 1.1   Distribution of nonzero elements in the HASM matrix 17

Fig. 2.1   Distribution of non-zero elements in the HASM.MOD matrix 47

Fig. 2.2   Eight mathematical surfaces. **a** f1. **b** f2. **c** f3. **d** f4. **e** f5. **f** f6. **g** f7. **h** f8 48

Fig. 2.3   Gaussian surface simulation. **a** HASM.MOD. **b** HASM 50

Fig. 2.4   Saddle surface simulation. **a** HASM.MOD. **b** HASM 50

Fig. 2.5   Comparison between the actual six fundamental coefficients of the Gaussian surface and the HASM simulated values. **a** E_real. **b** E_simu. **c** F_real. **d** F_simu. **e** G_real. **f** G_simu. **g** L_real. **h** L_simu. **i** M_real. **j** M_simu. **k** N_real. **l** N_simu 51

Fig. 2.6   Comparison between the actual six fundamental coefficients of the Gaussian surface and the HASM.MOD simulated values. **a** E_real. **b** E_simu. **c** F_real. **d** F_simu. **e** G_real. **f** G_simu. **g** L_real. **h** L_simu. **i** M_real. **j** M_simu. **k** N_real. **l** N_simu 52

Fig. 2.7   Comparison between the actual six fundamental coefficients of the saddle surface and the HASM simulated values. **a** E_real. **b** E_simu. **c** F_real. **d** F_simu. **e** G_real. **f** G_simu. **g** L_real. **h** L_simu. **i** M_real. **j** M_simu. **k** N_real. **l** N_simu 53

Fig. 2.8   Comparison between the actual six fundamental coefficients of the saddle surface and the HASM.MOD simulated values. **a** E_real. **b** E_simu. **c** F_real. **d** F_simu. **e** G_real. **f** G_simu. **g** L_real. **h** L_simu. **i** M_real. **j** M_simu. **k** N_real. **l** N_simu 54

Fig. 2.9   Distribution of meteorological stations in China 56

Fig. 2.10 Simulation of average temperature in January. **a** HASM. **b** HASM.MOD 59

Fig. 2.11 Simulation of average temperature in July. **a** HASM. **b** HASM.MOD 59

Fig. 2.12   Simulation of the average precipitation in January. **a**
            HASM. **b** HASM.MOD 61
Fig. 2.13   Simulation of the average precipitation in July. **a** HASM.
            b HASM.MOD 61
Fig. 2.14   Simulation areas of HASM and HASM.MOD 64
Fig. 2.15   Variation in GC (the left end of the Gauss-Codazzi
            equation; the number of grids is 2209) 68
Fig. 3.1   Simulation of the Gaussian surface using different driving
            fields: **a** HASM_0, **b** HASM_K, **c** HASM_I, **d** HASM_S 72
Fig. 3.2   Simulation of the saddle surface using different driving
            fields: **a** HASM_0, **b** HASM_K, **c** HASM_I, **d** HASM_S 73
Fig. 3.3   Simulation results of traditional HASM for eight surfaces
            with zero driving field: **a** f1, **b** f2, **c** f3, **d** f4, I f5, **f** f6, **g**
            f7, **h** f8 75
Fig. 3.4   Simulation results of HASM.MOD for eight surfaces
            with zero driving field: **a** f1, **b** f2, **c** f3, **d** f4, **e** f5, **f** f6, **g** f7,
            h f8 76
Fig. 3.5   Comparison of simulated and real temperature values
            in January: **a** HASM, **b** HASM.MOD 77
Fig. 3.6   Comparison of simulated and real temperature values
            in July: **a** HASM, **b** HASM.MOD 78
Fig. 3.7   Distribution of simulated temperatures in January:
            a HASM_K, **b** HASM_0, **c** HASM.MOD_K, **d**
            HASM.MOD_0 79
Fig. 3.8   Distribution of simulated temperatures in July:
            a HASM_K, **b** HASM_0, **c** HASM.MOD_K, **d**
            HASM.MOD_0 80
Fig. 3.9   Comparison of simulated and real precipitation in January:
            a HASM, **b** HASM.MOD 81
Fig. 3.10   Comparison of simulated and real precipitation in July: **a**
            HASM, **b** HASM.MOD 81
Fig. 3.11   National precipitation distribution in January: **a** HASM_K,
            b HASM_0, **c** HASM.MOD_K, **d** HASM.MOD_0 82
Fig. 3.12   National precipitation distribution in July: **a** HASM_K, **b**
            HASM_0, **c** HASM.MOD_K, **d** HASM.MOD_0 83
Fig. 4.1   Influence of the sampling ratio on the simulation results
            of the Gaussian surface 86
Fig. 4.2   Simulation of the Gaussian surface with different sampling
            ratios. **a** Sampling ratio of 1%. **b** Sampling ratio of 50% 87
Fig. 4.3   Error distribution between simulated and real Gaussian
            surface. **a** Sampling ratio of 1%. **b** Sampling ratio of 50% 87
Fig. 4.4   Influence of the sampling ratio on the simulation results
            of the saddle surface 88
Fig. 4.5   Simulation of the saddle surface at different sampling
            ratios. **a** Sampling ratio of 1%. **b** Sampling ratio of 50% 88

Fig. 4.6   Error distribution between the simulated and real values
            of the saddle surface. **a** Sampling ratio of 1%. **b** Sampling
            ratio of 50%   89
Fig. 4.7   Influence of the sampling ratio on the simulation results
            of surface f_1   89
Fig. 4.8   Simulation of surface f1 at different sampling ratios. **a**
            Sampling ratio of 1%. **b** Sampling ratio of 50%   90
Fig. 4.9   Error distribution between the simulated and real values
            of surface f_1. **a** Sampling ratio of 1%. **b** Sampling ratio
            of 50%   90
Fig. 4.10   Influence of the sampling ratio on the simulation results
            of average temperature in January   91
Fig. 4.11   HASM.MOD simulation of January temperature
            at different sampling ratios. **a** Sampling ratio of 30%. **b**
            Sampling ratio of 90%   91
Fig. 4.12   Differences between simulated and real temperatures
            in January at different sampling ratios   92
Fig. 4.13   Influence of the sampling ratio on the simulation results
            of the average temperature in July   92
Fig. 4.14   Simulation of July temperature by HASM.MOD
            at different sampling ratios. **a** A sampling ratio of 30%. **b**
            A sampling ratio of 90%   93
Fig. 4.15   Difference between simulated and measured values of July
            temperatures at different sampling ratios   93
Fig. 4.16   Influence of the sampling ratio on the simulation results
            of precipitation in January   94
Fig. 4.17   Simulation of precipitation in January by HASM.MOD
            with different sampling ratios. **a** A sampling ratio of 30%.
            b A sampling ratio of 90%   94
Fig. 4.18   Differences between simulated and measured values
            of precipitation in January at different sampling ratios   95
Fig. 4.19   Influence of the sampling ratio on the simulation results
            of precipitation in July   95
Fig. 4.20   Simulation of precipitation in July by HASM.MOD
            at different sampling ratios. **a** A sampling ratio of 30%. **b**
            A sampling ratio of 90%   96
Fig. 4.21   Differences between the simulated and measured values
            of precipitation in July at different sampling ratios   96
Fig. 4.22   Influence of the sampling error on the simulation results
            of the Gaussian surface   98
Fig. 4.23   Simulation of the Gaussian surface at different sampling
            errors. **a** A sampling error of 0.0000577. **b** A sampling
            error of 0.0577   98

Fig. 4.24   Differences between the simulated and real Gaussian
             surfaces at different sampling errors. **a** A sampling error
             of 0.0000577. **b** A sampling error of 0.0577   99
Fig. 4.25   Influence of the sampling error on the simulation accuracy
             of the saddle surface   99
Fig. 4.26   Simulation of the saddle surface at different sampling
             errors. **a** A sampling error of 0.00000577. **b** A sampling
             error of 0.0577   100
Fig. 4.27   Differences between the simulated and real saddle
             surfaces at different sampling errors. **a** A sampling error
             of 0.00000577. **b** A sampling error of 0.0577   100
Fig. 4.28   Influence of the sampling error on the simulation results
             of surface f_1   101
Fig. 4.29   Simulation of surface f_1 at different sampling errors. **a**
             A sampling error of 0.00000577. **b** A sampling error
             of 0.0577   101
Fig. 4.30   Differences between the simulated and the real values
             of surface f_1 with different sampling errors. **a** A sampling
             error of 0.00000577. **b** A sampling error of 0.0577   101
Fig. 4.31   Influence of the sampling error on the simulation results
             of average temperature in January   102
Fig. 4.32   Simulation of the average temperature in January
             with different sampling errors. **a** A sampling error
             of 0.00000288. **b** A sampling error of 2.87   102
Fig. 4.33   Differences between the simulated and measured
             temperature in January with different sampling errors   103
Fig. 4.34   Influence of the sampling error on the average temperature
             in July   103
Fig. 4.35   Simulation results of temperature in July with different
             sampling errors. **a** A sampling error of 0.0000288. **b**
             A sampling error of 2.87   104
Fig. 4.36   Differences between the simulated and measured values
             of temperature in July at different sampling errors   105
Fig. 4.37   Influence of the sampling error on the simulation results
             of precipitation in January   105
Fig. 4.38   Simulation of precipitation in January with different
             sampling errors. **a** A sampling error of 0.0000288. **b**
             A sampling error of 2.87   106
Fig. 4.39   Differences between the simulated and measured values
             of precipitation in January at different sampling errors   106
Fig. 4.40   Influence of the sampling error on the simulation results
             of precipitation in July   107
Fig. 4.41   Simulation of the precipitation in January with different
             sampling errors. **a** A sampling error of 0.00000288. **b**
             A sampling error of 2.87   107

Fig. 4.42   Differences between the simulated and measured values
            of precipitation in July at different sampling errors 108
Fig. 5.1   Convergence diagram of the CG method. **a** A small K
            value. **b** A large K value 113
Fig. 5.2   Comparison of the computation times of different
            preconditioning methods 117
Fig. 5.3   Real (**a**) and simulated (**b**) Gaussian surfaces 122
Fig. 5.4   Computation time and speed-up ratio of the HASM.MOD
            parallel algorithm 122
Fig. 6.1   Spatial distribution of the meteorological stations in China
            and four subareas 127
Fig. 6.2   Average seasonal precipitation from 1951 to 2010 in each
            subarea 135
Fig. 6.3   Spatial distribution of the average seasonal precipitation
            between 1951 and 2010 in China (from the upper left
            to lower right: spring, summer, autumn, and winter) 136
Fig. 7.1   Location of meteorological stations in China 141
Fig. 7.2   Comparison between HASM.MOD and HASM. **a**
            Maximum. **b** Minimum. **c** Mean 142
Fig. 7.3   Distributions of the percentage of sunshine obtained
            by HASM.MOD. **a** January. **b** April. **c** July. **d** October 143
Fig. 8.1   Location of the meteorological stations in the Heihe River
            Basin 149
Fig. 8.2   Comparison of average monthly ET data of Mazongshan
            station 160
Fig. 8.3   Comparison of average monthly ET data of Yumen Town
            station 161
Fig. 8.4   Comparison of average monthly ET data of Jiuquan station 162
Fig. 8.5   Comparison of average monthly ET data of Zhangye station ... 162
Fig. 8.6   Simulation results of potential ET in the Heihe River Basin
            in January 163
Fig. 8.7   Simulation results of potential ET in the Heihe River Basin
            in July 164
Fig. 9.1   Spatial distribution of the meteorological stations
            and DEM in the Heihe River Basin 172
Fig. 9.2   Comparison between the average temperatures
            of the observation values and CMIP5 results
            and the CMIP5 results after downscaling 176
Fig. 9.3   Comparison between the average precipitation observation
            values and CMIP5 results and the CMIP5 results
            after downscaling 176
Fig. 9.4   Distributions of temperature for the CMIP5 results
            and downscaled CMIP5 results 177
Fig. 9.5   Distributions of precipitation for CMIP5 results
            and downscaled CMIP5 results 178

Fig. 9.6   Downscaled results of temperature in different periods
              under the RCP2.6 scenario   180
Fig. 9.7   Downscaled results of temperature in different periods
              under the RCP4.5 scenario   181
Fig. 9.8   Downscaled results of temperature in different periods
              under the RCP8.5 scenario   182
Fig. 9.9   Downscaled results of precipitation in different periods
              under the RCP2.6 scenario   183
Fig. 9.10   Downscaled results of precipitation in different periods
              under the RCP4.5 scenario   184
Fig. 9.11   Downscaled results of precipitation in different periods
              under the RCP8.5 scenario   185

List of Tables

Table 2.1   Calculation errors of HASM and HASM.MOD   49

Table 2.2   Differences between simulation results of HASM
and HASM.MOD and real values   49

Table 2.3   Simulation accuracies of HASM and HASM.MOD
for monthly average temperature   58

Table 2.4   Simulation of monthly average precipitation   61

Table 2.5   Computational errors (RMSE) of different discretization
schemes of the three Gaussian equations using different
numbers of grids   62

Table 2.6   Computational errors of HASM and HASM.MOD
at the boundaries (number of grids: 441)   65

Table 2.7   Number of outer iterations upon convergence
of HASM.MOD   67

Table 2.8   Errors (RMSE) of HASM.MOD under different
convergence criteria for outer iteration   67

Table 3.1   Simulation error of traditional HASM for the Gaussian
surface with different driving fields   72

Table 3.2   Simulation error of traditional HASM for the saddle
surface with different driving fields   73

Table 3.3   Simulation errors (RMSE) of HASM and HASM.MOD
with zero driving field   74

Table 3.4   Differences between HASM and HASM.MOD regarding
temperature simulation under different driving fields   77

Table 3.5   Differences in simulated precipitation of HASM
and HASM.MOD with different driving fields   80

Table 5.1   Simulation accuracy of different methods with different
numbers of inner iterations   118

Table 5.2   Simulation accuracies of different methods with different
numbers of outer iterations   118

Table 6.1   Area and number of stations in each subarea   128

Table 6.2   Local topographic factors influencing precipitation   129

Table 6.3   Geographical and topographic factors influencing
             precipitation in each subarea during different seasons
             and the R^2 of the corresponding regression equation 130
Table 6.4   RMSE, MAE, MRE for HASM, kriging, IDW and spline
             results (summer) 133
Table 6.5   RMSE, MAE, MRE for HASM, kriging, IDW and spline
             results (winter) 134
Table 7.1   Comparison of the accuracies of different methods 141
Table 8.1   Average monthly ET data in the Heihe River Basin
             in 2000–2009 152
Table 8.2   HASM error analysis (unit: mm) 153
Table 8.3   Error analysis of the meteorological forcing data (unit: mm) ... 154
Table 8.4   Error analysis of the kriging method (unit: mm) 155
Table 8.5   Error analysis of the IDW method (unit: mm) 156
Table 8.6   Error analysis of the spline method (unit: mm) 157
Table 8.7   Comparison of the monthly potential ET results
             from the six methods (unit: mm) 165
Table 9.1   Comparison of the CMIP5 output and downscaling results 175
Table 9.2   Comparison of different downscaling methods for CMIP5
             temperature data 178
Table 9.3   Comparison of different downscaling methods for CMIP5
             precipitation data 178

Chapter 1
Introduction

1.1 Research Background and Significance

1.1.1 Surface Modelling

Studying the spatial variation in a specific variable is often required to solve many practical problems, yet in most cases, the values of only some sparse and disordered discrete points can be obtained due to limitations. To obtain the attribute value of any point on the whole surface, the values of unknown points must be calculated per certain rules based on these known discrete points. Surface reconstruction refers to the process of constructing an original surface based on the information of several sampling points on the surface. Surface modelling is a key step in reverse engineering, through which discrete observation data can be reconstructed into a continuously changing surface. According to different approaches of surface construction, surface modelling can be divided into two categories, i.e., surface interpolation and surface approximation. In surface interpolation, the simulated area is interpolated per the sampling points according to a certain mathematical description form, and the interpolated surface should strictly pass through the control points. In surface approximation, a grid surface is constructed based on the discrete data. The grid is used to approach the surface to bring the surface close to the control points within a certain error control range. Currently, computers are widely used to achieve this. For practical applications in aviation, shipbuilding, and precision machining, surface reconstruction has been further developed because it is particularly important both practically and theoretically.

Surface modelling is very important in geoscience research. Spatial surface simulation of data is common in geoscience research as well as resource and environmental investigation. Many practical problems should be further analysed based on data with a complete spatial distribution. Geoscience research usually involves massive amounts of data, some complete and some incomplete, which is particularly true for data obtained by meteorological stations. Although the total number of meteorological stations has increased, the number of meteorological stations within a specific

© The Author(s), under exclusive license to Springer Nature Singapore Pte Ltd. 2021   1
N. Zhao and T. Yue, *High Accuracy Surface Modeling Method: The Robustness*,
https://doi.org/10.1007/978-981-16-4027-8_1

study area is usually still low. There are no meteorological stations distributed in a wide range of areas, making it impossible to obtain the true value of any sampling point; sometimes, there are certain data in a study area, but the density of data acquisition cannot meet the requirements of further analysis. The surface modelling method can construct a continuous surface based on discrete data. The number of meteorological stations is limited in studies of the spatial distributions of meteoro-logical elements. With effective observation data, the optimized surface modelling method is not only a prerequisite for achieving high accuracy and efficient simu-lation of the spatial distribution of meteorological elements but also an important research field in meteorological interpolation. However, due to the different data sources and diverse distribution of sampling points, the method used directly deter-mines the final accuracy of the simulated surface (Maceachren and Davidson 1987; Rhind 1975). Inappropriate surface modelling methods can generate incorrect results, leading to incorrect conclusions for practical problems. Especially for areas with great terrain changes in China, how to choose an appropriate method has become an urgent problem. Hence, exploring the surface modelling methods and improving current high-accuracy surface modelling (HASM) theories are very important for geoscience research.

1.1.2 HASM of Climatic Elements

Temperature and precipitation are key indicators of regional and even global climate systems, and changes in the spatial distribution of these key indicators directly affect industrial and agricultural production and therefore influence the development of the national economy.

Meteorological elements serve as the basic data for studies of geography, agri-culture, ecology, and global change, which play an especially important role in the simulation of regional and global changes in the ecosystem and the study of ecosystem management. The biogeochemical cycling system is particularly signifi-cant to human beings and nature, and climate change could alter it by affecting land use patterns, modifying hydrological cycling systems, and eliminating biodiversity (Dobson et al. 1997). Varying climatic conditions cause spatiotemporal changes in the spatial pattern and functional structure of an ecosystem, which greatly influences social and economic systems. Moreover, with the aggravation of climate change and the frequent occurrence of abnormal climate events, there could definitely be a more serious impact. Temperature and precipitation are the most important climatic elements. Changes in the spatiotemporal distribution of temperature and precipitation could directly change the structure, function, and spatial distribution of ecosystems. Driven by a wave of worldwide research on climate change, many models reflecting the evolution pattern of ecosystems have emerged (Zhao et al. 2008), including the soil-vegetation-atmospheric transmission (SVAT) model (Roni 1998; Liu and Sun 1999), models of regional vegetation productivity and carbon cycling (the Carnegie-Ames-Stanford-approach (CASA) model by Hanqin et al. (1998) and Hasenauer et al.

(2003), the trajectory equifinality model (TEM) by Sato et al. (2004), and the forest-biogeochemistry (FOREST-BGC) model by Band 1991), vegetation dynamic models (the Lund-Potsdam-Jena general ecosystem simulator (LPJ-GUESS) by Smith et al. 2001), and hydrological models (the soil water assessment tool (SWAT) by Tong and Naramngam 2007). All these models require high-accuracy temperature and precipitation data as the input parameters. In hydrological research, precipitation is an input term for studying various hydrological processes on watersheds, and the quality of precipitation simulation significantly affects the simulation accuracy of distributed hydrological models. High-accuracy data on the spatial distribution of precipitation is required by distributed hydrological models to explore various hydrological processes and water transformation in watersheds (Fan and Bras 1995; Teegavarapu 2009). However, it is difficult to directly obtain these high-accuracy environmental factors from remote sensing images. Instead, these factors should be generated in geographic information systems (GISs) using interpolation methods based on station data. High-quality and high-accuracy data on the spatial distributions of temperature and precipitation are not only the basis for the comprehensive evaluation of changes in ecosystems but also the prerequisites for the effective operation of many ecosystem models, e.g., ecosystem process models, ecosystem pattern models, and biological community models (Fang 2002; Yue et al. 2005; Yue et al. 2007).

However, the observation data obtained at meteorological stations are only local and discrete data from limited spatial points. To meet the requirements of precision agriculture, global change science, and ecology, HASM methods must be developed to obtain continuous and orderly spatial data. The data at the unknown points are estimated by interpolation based on the observation data obtained at meteorological stations, which has been a popular research topic in geographical science ever since the twentieth century. In addition, with the development of ecology and research on global change, the spatial distribution of meteorological elements with high spatiotemporal resolution is urgently needed. Therefore, researchers have focused on how to estimate the values of climatic elements in areas without observation points based on the spatial distribution patterns of climatic elements.

1.2 Research Background

1.2.1 Common Surface Reconstruction Methods

1.2.1.1 Curvature Interpolation Methods

Global Interpolation

Global interpolation methods establish fitting models based on the observation values of all the sampling points in the area of interest. Generally, global interpolation is achieved by polynomial functions. It is often used to simulate macro-trends in a wide range and cannot provide the local characteristics of the interpolated region. Theoretically, any complex surface can be approximated by high-order polynomials with any arbitrary precision. However, the coefficients of polynomials do not have clear physical significance, oscillations can easily occur, and the computational cost is high; therefore, global interpolation is not widely applied.

Local Interpolation

Due to the complexity and variation in the curvature of actual surfaces, it is often impossible to fit an entire area with merely one polynomial function, and the results of global interpolation are often not ideal. Therefore, the modelled surface is divided into several sub-areas, and each area is then interpolated using different models. This method can more effectively reflect the local surface features.

(1) Linear Interpolation

Linear interpolation determines a plane based on three known points neighbouring the interpolation point; this plane is then used to calculate the value of the interpolation point. This simple method is extensively used in triangulated irregular network (TIN) interpolation. The corresponding function is expressed as follows:

$$Z = a_0 + a_1 x + a_2 y$$

where a_0, a_1, and a_2 are undetermined coefficients and x, y, and z denote the three-dimensional coordinates of the data point. However, if the three reference points are approximately distributed near a straight line, the solution obtained by this method is unstable.

(2) Bilinear Interpolation

Bilinear interpolation determines a bilinear polynomial function using four known data points neighbouring the interpolation point; this plane is then used to determine the value of the interpolation point. The corresponding function is expressed as follows:

$$Z = a_0 + a_1 x + a_2 y + a_3 xy$$

A cross term not present in linear interpolation is added in bilinear interpolation. This method is currently widely applied to grid-based interpolation.

(3) Hardy's Multiquadric (MQ) Interpolation

Hardy's MQ interpolation was proposed by Professor Hardy in 1971 (Hardy 1971). In Hardy's MQ interpolation based on numerical approximation, topography surfaces are approximated using polynomial and Fourier series. The basic idea is that any continuous surface can be approximated by several simple surfaces. With the poly-hedral function as the interpolation function, several simple surfaces are superim-posed to represent the curved surface, thereby achieving a better fitting result. The corresponding function is expressed as follows:

$$Z_i = \sum_{i=1}^{n} k_i Q(x, y, x_i, y_i)$$

where $Q(x, y, x_i, y_i)$ is the selected mathematical surface, i.e., the kernel func-tion, which is generally a symmetric function with a simple form; n denotes the number of layers of the superimposed surfaces; k_i refers to an undetermined param-eter, which represents the contribution of the i-th kernel function to the polyhedral function. The Hardy's interpolation equation is solved by determining several unde-termined coefficients $k_i (i = 1, 2, \ldots, n)$ under the least squares (LS) constraint. This simple and fast interpolation method allows convenient addition of constraints and restrictions to the surface. However, the method has extensive uncertainty in regard to the selection of the kernel function, and different kernel functions may greatly affect the results.

(4) Spline Interpolation

Spline interpolation is a method that uses piecewise polynomials to fit a surface (Wahba and Wendelberger 1980; Eckstein 1989; Hutchinson and Gessler 1994). It uses mathematical methods to generate a smooth curve that passes through the known sampling points. When early engineers designed drawing, they fixed a slender elastic strip (spline) at the sampling points, and since the remaining strip bent freely, curves could be drawn to form splines (Li et al. 2006). The calculation of spline interpolation is based on the concept of minimum curvature (Hartkamp et al. 1999). Usually, the interpolation area is first divided into several sub-areas, and each subarea is then defined with a polynomial function. To ensure that the surfaces are smooth, the n-1-order derivatives of the determined n-order polynomial surface should be continuous at the boundaries with all adjacent sub-areas. Spline interpolation is simple and has a relatively low computational cost, and there is no need to make statistical assumptions about the interpolation data in advance; thus, it is a relatively robust method applicable to smooth surfaces. The limitation of spline interpolation is that it is difficult to estimate the error, and the interpolation result is not ideal when there are limited sampling points or when the attribute value changes drastically within a very short distance.

(5) Kriging Interpolation

Although classical interpolation theory has been thoroughly developed, spatial corre-lation should be considered when applying this method to geoscience to improve the interpolation accuracy. The validity and advantages of spatial correlation have been effectively verified both theoretically and practically, and a certain mechanism has been integrated with mathematical analysis. Spatial correlation theory, as a major feature of spatial data analysis and prediction, turns some problems in geoscience into branches of mathematical problems with their own features. Classical mathematical theory still serves as the foundation of spatial data analysis, yet the theory should be integrated with practical problems and spatial correlation to obtain more accurate and reliable results.

The kriging interpolation method is a linear regression method and is also known as the geostatistical method. It was first proposed by Krige in 1951 and has been gradually developed since the creation of geostatistics by Matheron and his research team (Matheron 1963). The kriging method is based on regionalized variable theory (Journel and Huijbregts 1978; Goovaerts 1997, 1999a) and has attracted extensive attention because it considers the spatial correlation between variables (Courault and Monestiez 1999; Joly et al. 2011). As an LS regression algorithm, kriging interpolation aims to reduce the expected error to zero while minimizing the variance (Hou and Huang 1990). The estimated value $Z(x)$ of the variable at x is expressed by the following equation:

$$Z(x) = \sum_{i=1}^{n} \lambda_i Z(x_i)$$

.

where λ_i denotes the weighting coefficient. Since the kriging method is based on the assumption that the data are spatially stable, $Z(x)$ is assumed to be intrinsically smooth and steady. In other words, if the variances between any two points with the same distance and direction are the same, then the weighting coefficient λ_i satisfies the following kriging equation set:

$$\begin{cases} \sum_{j=1}^{n} \lambda_j \gamma(x_i, x_j) + \mu = r(x_i, x), i = 1, \cdots, n \\ \sum_{j=1}^{n} \lambda_j = 1 \end{cases},$$

where $\gamma(x_i, x_j)$ denotes the variogram value between sampling points x_i and x_j and μ is the Lagrangian constant. The aim of the kriging method is to obtain the optimal variogram equation. The estimation of parameters is the key to the selection of the theoretical model of the variogram. The selection of the model and its parameters is a process of multi-parameter non-linear optimization. Since the variogram usually includes discontinuous derivatives, the estimation of the parameters is complicated (Zeng and Huang 2007).

After years of improvement, the kriging method has evolved into many forms, e.g., the ordinary kriging method, the universal kriging method, and the co-kriging method, that can be used in different scenarios. Note that the premise of the kriging method is that the data should be spatially stable; this premise cannot easily be satisfied in practical application. In addition, the selection of the variogram is a challenge.

Pointwise Interpolation

With the interpolation point as the centre, a local function is used to fit the surrounding data points; the range of the data points varies with the position of the interpolation point.

(1) Moving Method

For each interpolation point, several adjacent data points are selected and fitted with a polynomial surface, which is usually a quadratic polynomial.

$$Z = a_0 + a_1 x + a_2 y + a_3 x^2 + a_4 y^2 + a_5 xy$$

The key to this method is to determine the minimum neighbourhood range of the interpolation point to ensure sufficient sampling points. In addition, the weights of the neighbouring points should be determined. According to the first law of geography, the closer the observation points are, the more similar these points are to each other, and vice versa. For the moving fitting method, the weight function is usually related to the distance.

(2) Weighted Average Method

The moving method needs to solve complex error equations and is often replaced by weighted the average method in practical application. When the value of a point is determined, the weighted average method replaces the error equation with the weighted average. Inverse distance weighting (IDW), as a special weighted average method, is widely applied in a range of fields. As a type of distance weighted coefficient method, IDW is similar to the nearest neighbour method; the principle of this method is that the weight given to a nearby point is greater than that given to a distant point. IDW was first proposed by Shepard and has been subsequently modified. According to the orders of distance used to calculate the weight, IDW can be further divided into inverse distance, inverse-square-distance, and inverse-high-order-distance. The IDW method assumes that the value of the interpolation point is the weighted average of the known points within a certain neighbourhood around the point. Let $Z^*(x)$ be the estimated value of the regionalized variable Z at unknown point x, then

$$Z^*(x) = \sum_{i=1}^{n} w_i Z(x_i)$$

where $x_i, i = 1, \cdots, n$ denote the sampling points within a certain neighbourhood around the unknown point x, n is the number of sampling points in this neighbourhood, w_i represents the weight of each sampling point, $w_i = \frac{1/d_i^p}{\sum_{i=1}^{n} 1/d_i^p}$, d_i is the distance between the point to be estimated and the sampling point, and p is the exponential parameter.

The IDW method calculates the weighted spatial distance of each sampling point. When the weight is 1, the interpolation is linear; when the weight is greater than 1, the interpolation is non-linear. The IDW method is simple and easy to implement, and the interpolation results can be adjusted by modifying the weight coefficients. However, the method considers only the distance information between spatial data, while ignoring a series of spatial attribute information such as spatial correlation. Moreover, IDW is sensitive to the choice of the weight function and is greatly affected by a non-uniform distribution of data points. Another shortcoming is the bulls eye effect, which describes concentric areas of the same value around known data points.

There are a variety of curvature interpolation methods, which can be divided into various types per different classification standards. According to whether all the data points are used during interpolation, methods can be divided into global interpolation and local interpolation. In global interpolation, one point is predicted based on the information of multiple data points, and the surface is thus relatively smooth. All the data are used in global interpolation, and a single interpolation point could be affected by the data within the whole region. Hence, once the value of a single data point is changed, the interpolation results of the whole region could be affected. In contrast, local interpolation considers only the adjacent points around the interpolation point; the changes in a single data point influence only a limited area around the interpolation point. Moreover, global interpolation can be localized by adjusting parameters, such as the radius and number of points used in the calculation. When a single model is not suitable for a large dataset, local interpolation is more appropriate. In a study by He et al., interpolation is divided into global interpolation, local interpolation, and hybrid interpolation. As a combination of global interpolation and local interpolation, hybrid interpolation further increases the interpolation accuracy by correcting the residual error in global interpolation. The accuracy of global interpolation has an important influence on hybrid interpolation, and global interpolation is an inaccurate interpolation method. Combining the strengths of global interpolation and local interpolation, hybrid interpolation is an effective interpolation method. According to the simulation value of an interpolated sampling point, interpolation can be divided into deterministic interpolation and random interpolation. In deterministic interpolation, the predicted value at the sampling point is equal to the measurement value at the sampling point. However, this constraint can be relaxed if the measured error is very large. The classical methods commonly used for spatial interpolation of meteorological elements include the IDW method, the spline method, and the kriging method.

1.2.1.2 Surface Approximation

Establishing a mathematical model is only the first step of applying mathematics to practical problems. The many mathematical models, in essence, are about solving differential equations. Differential equations are the bridge between mathematical knowledge and practical applications. To apply partial differential equations to simulate surfaces, it is necessary to first discretize the differential equations and then solve the algebraic system. The minimum curvature method is representative of surface grid approximation and has been integrated into the software Surfer.

Minimum Curvature

Minimum curvature is often used in surface estimation in geoscience. The surface generated using minimum curvature is a long, thin splinter with the least amount of bending that passes through the known points to the maximum extent (Weng 2006). Through the control of the convergence standard of the minimum curvature through the parameters of the maximum cycle coefficient and the maximum residual error, the minimum curvature method is expressed as:

$$C(u)_{min} = \iint \left(\frac{\partial^2 u}{\partial x^2} + \frac{\partial^2 u}{\partial y^2} \right)^2 dx\,dy$$

Studies have shown that low values are often smoothed in areas with large slope variation and that the time complexity of the above equation is proportional to the cubic power of the number of grids in the simulation area.

1.2.2 High Accuracy Surface Modelling (HASM)

1.2.2.1 Background of HASM

Whether an attribute value of a point is to be estimated or an isoline or grid surface is to be generated from irregular data at observation points, spatial interpolation fits a function based on the attribute value and spatial position of the given discrete point, and this function fully reflects the mathematical relationship between the attribute value and spatial position of the given point, which can help with the inference of the attribute values of other points within a specific area. In essence, spatial interpolation reconstructs a function that fully approximates the spatial distribution characteristics of elements through modelling.

Previous surface modelling methods have been based either on geostatistical theory and the neighbourhood correlation hypothesis or the elastic mechanism rather than a comprehensive consideration of the surface elements. Since the constraint effect of the intrinsic factors on surface reconstruction is not considered in modelling, it is difficult to control the error of the surface fitting model. HASM is a new method that has been developed in recent years that can be used for spatial interpolation and prediction (Yue 2011). Based on a study by Evans (1980), the slope surface, aspect and curvature are important variables reflecting the local characteristics of curved surfaces. Yue and Ai (1990) were the first to establish a cirque-shaped mathematical model based on curve theory; this model was used to detect the changes in the surface of Earth (Yue et al. 2002). According to differential geometry, a spatial curve is determined by its curvature and torsion, and a spatial surface is determined by the first and second fundamental coefficients of the surface (Okubo 1987; Toponogov 2006). Therefore, the surface modelling technology should focus on the determinants of the surface itself. On this basis, Yue et al. developed the HASM method. The theoretical system of HASM has been greatly improved ever since 2004, and the error problem associated with classical surface modelling in geographic information systems (GISs) has been solved (Yue et al. 2004; Yue and Du 2005, 2006a, b; Yue et al. 2007; Yue and Du et al. 2007a; Yue and Li 2010a,b; Yue et al. 2010; Yue et al. 2012). Both numerical simulation and experimental verification have shown that the simulation accuracy of HASM is higher than that of classical interpolation methods (Yue and Du et al. 2007b; Yue and Song 2008; Yue et al. 2008; Shi et al. 2009; Yue and Li 2010; Yue and Wang 2010; Yue et al. 2010a,b; Yue et al. 2013a, b; Zhao and Yue 2014).

1.2.2.2 Theoretical Basis of HASM

Although Evans (1980) points out that slope surface, aspect and curvature are important variables in the local characteristics of a surface, according to differential geometry, the slope surface, aspect and curvature are only the determinants of the hatching lines of the surface, while the surface is determined by the first and second fundamental forms of the surface (Toponogov 2006).

First Fundamental Form of a Surface

Let $P(x, y)$ and $P'(x + \Delta x, y + \Delta y)$ be 2 adjacent points on surface S and the surface S be $r = (x, y, f(x, y))$, then

$$PP' = r(x + \Delta x, y + \Delta y) - r(x, y) = r_x \Delta r_y \Delta y + \cdots$$

When P is infinitely close to P', the higher-order term of Δx and Δy (> the second order) can be omitted, and $dr = r_x dx + r_y dy$. In this case, ds, the main part of the length of vector PP', is defined as the distance between the two infinite adjacent points on surface S:

$$ds^2 = |dr|^2 = dr \cdot dr = (r_x dx + r_y dy) \cdot (r_x dx + r_y dy)$$
$$= r_x \cdot r_x (dx)^2 + 2r_x \cdot r_y \cdot dx \cdot dy + r_y \cdot r_y (dy)^2$$

$E = r_x \cdot r_x$, $F = r_x \cdot r_y$, $G = r_y \cdot r_y$. The above equation can be rewritten as:

$$I = ds^2 = E dx^2 + 2F dx dy + G dy^2 \tag{1.1}$$

Equation (1.1) is the first fundamental form of the surface, and E, F and G are the coefficients of the surface, also known as the first fundamental coefficients. Since $dr = r_x dx + r_y dy$, dr is a vector in the tangent plane of the surface, and $ds^2 = dr \cdot dr$ is a positive definite quadratic form, $E > 0$, $G > 0$, $and\, EG - F^2 > 0$. The geometric quantities expressed by the coefficient of the first fundamental form of the surface are considered to be the intrinsic quantities, e.g., the length of the curve on the surface, the angle between two curves, the area of a specific region on the surface, the geodesic line, and the curvature and total curvature of the geodesic line. The intrinsic quantities remain unchanged when the surface is deformed.

Second Fundamental Form of a Surface

To study the degree of bending at point P on the surface, the vertical distance δ from the point P' (near point P) to the tangent plane of P is calculated.

$$PP' = r(x + \Delta x, y + \Delta y) - r(x, y) = r_x \Delta x + r_y \Delta y + \frac{1}{2}(r_{xx}(\Delta x)^2 + 2r_{xy}\Delta x \Delta y$$

$+ r_{yy}(\Delta y)^2) + \cdots$, and the higher-order terms of Δx and Δy($>$ the third order) are omitted, then

$$\delta = PP' \Delta n = \frac{1}{2}\left(r_{xx}\Delta n(\Delta x)^2 + 2r_{xy}\Delta n \Delta x \Delta y + r_{yy}\Delta n(\Delta y)^2\right) + \cdots$$

n is the normal vector of S. Let $L = r_{xx} \cdot n$, $M = r_{xy} \cdot n$, $N = r_{yy} \cdot n$, then

$$2\delta = L(\Delta x)^2 + 2M\Delta x \Delta y + N(\Delta y)^2 + \cdots$$

When P' approaches P, the main part of 2δ is

$$II = Ldx^2 + 2Mdxdy + Ndy^2 \tag{1.2}$$

which is considered to be the second fundamental form of a surface, where L, M, N are the coefficients of the second fundamental form and are also known as the second fundamental coefficients.

Since $r_x \cdot n = 0$, $r_y \cdot n = 0$, the derivatives with respect to x and y are taken on both sides,

$$r_{xx} \cdot n + r_x \cdot n_x = 0, r_{xy} \cdot n + r_x \cdot n_y = 0,$$

$$r_{yx} \cdot n + r_y \cdot n_x = 0, r_{yy} \cdot n + r_y \cdot n_y = 0,$$

Thus, $L = -r_x \cdot n_x$, $M = -r_x \cdot n_y$, $N = -r_y \cdot n_y$. The second fundamental form is written as

$$II = Ldx^2 + 2Mdxdy + Ndy^2 = -dr \cdot dn = -(dn, dr).$$

The second fundamental form describes the shape of the surface and reflects the local curvature variation of the surface.

Principal Theorem of the Surface Theory

Based on the principal theorem of the surface theory (Su and Hu 1979), let the first and second fundamental coefficients of the surface E, F, G, L, M, N be symmetrical, let E, F, G be positive definite, and let E, F, G, L, M, N satisfy the Gauss-Codazii equation set, then $z = f(x, y)$ is the unique solution to the total differential equation set (1.3) under the initial condition of $f(x, y) = f(x_0, y_0)(x = x_0, y = y_0)$.

$$
\begin{cases}
f_{xx} = \Gamma_{11}^1 f_x + \Gamma_{11}^2 f_y + \dfrac{L}{\sqrt{E + G - 1}} \\[2mm]
f_{yy} = \Gamma_{22}^1 f_x + \Gamma_{22}^2 f_y + \dfrac{N}{\sqrt{E + G - 1}} \\[2mm]
f_{xy} = \Gamma_{12}^1 f_x + \Gamma_{12}^2 f_y + \dfrac{M}{\sqrt{E + G - 1}} \\[2mm]
\dfrac{f_y f_{xy} + f_x f_{xx}}{\left(1 + f_x^2 + f_y^2\right)^{3/2}} = L\Gamma_{12}^1 + M\left(\Gamma_{12}^2 - \Gamma_{11}^1\right) - N\Gamma_{11}^2 \\[2mm]
\dfrac{f_x f_{xy} + f_y f_{yy}}{\left(1 + f_x^2 + f_y^2\right)^{3/2}} = L\Gamma_{22}^1 + M\left(\Gamma_{22}^2 - \Gamma_{12}^1\right) - N\Gamma_{12}^2
\end{cases}
\tag{1.3}
$$

where $E = 1 + f_x^2, F = f_x \cdot f_y, G = 1 + f_y^2,$

$$L = \frac{f_{xx}}{\sqrt{1 + f_x^2 + f_y^2}}, M = \frac{f_{xy}}{\sqrt{1 + f_x^2 + f_y^2}}, N = \frac{f_{yy}}{\sqrt{1 + f_x^2 + f_y^2}},$$

$$\Gamma_{11}^1 = \frac{1}{2}(GE_x - 2FF_x + FE_y)(EG - F^2)^{-1},$$

$$\Gamma_{11}^2 = \frac{1}{2}(EF_x - EE_y - FE_x)(EG - F^2)^{-1},$$

$$\Gamma_{22}^1 = \frac{1}{2}(2GF_y - GG_x - FG_y)(EG - F^2)^{-1},$$

$$\Gamma_{22}^2 = \frac{1}{2}(EG_y - 2FF_y + FG_x)(EG - F^2)^{-1},$$

$$\Gamma_{12}^1 = \frac{1}{2}(GE_y - FG_x)(EG - F^2)^{-1},$$

$$\Gamma_{12}^2 = \frac{1}{2}(EG_x - FE_y)\left(EG - F^2\right)^{-1}.$$

The Gauss-Codazii equation set is

$$\begin{cases} \left(\frac{L}{\sqrt{E}}\right)_y - \left(\frac{M}{\sqrt{E}}\right)_x - N\frac{\sqrt{E_y}}{G} - M\frac{\sqrt{G_x}}{\sqrt{EG}} = 0 \\ \left(\frac{N}{\sqrt{G}}\right)_x - \left(\frac{M}{\sqrt{G}}\right)_y - L\frac{\sqrt{G_x}}{E} - M\frac{\sqrt{E_y}}{\sqrt{EG}} = 0 \\ \left(\frac{\sqrt{E_y}}{\sqrt{G}}\right)_y + \left(\frac{\sqrt{G_x}}{\sqrt{E}}\right)_x + \frac{LN-M^2}{\sqrt{EG}} = 0 \end{cases} \tag{1.4}$$

Note: 1. The first three equations in the differential equation set (1.3) are partial differential equations of the surface, which are also known as the Gauss equation set, and the latter two equations are referred to as the Weingarten equation set. Somasundaram (2005) pointed out that the Weingarten equation set is a complement to the Gauss equation set, while the Gauss equation set could be regarded as the partial differential equation set of the surface.

2. When $f_x(x, y)$, $f_y(x, y)$, $n_x(x, y)$, and $n_y(x, y)$ (n is the normal vector of surface $z = f(x, y)$) satisfy the condition that the second partial derivatives of x and y are commutative, the Gauss-Codazii equation set could be derived from the Gauss-Weingarten equation set (Su and Hu 1979).

According to the principal theorem of surface theory, if the first and second fundamental coefficients of the surface are symmetrical and positive definite and satisfy the Gauss-Codazii equation set, the surface can be determined by the partial differential equation set of the surface, i.e., the Gauss equation set. Since HASM is based on the fundamental theorem of surface theory, the partial differential equations of the surface should be first discretized into an algebraic equation set and then solved by the iteration method.

According to its development stage, HASM can be divided into HASM1 (Yue 2011), HASM2 (Yue et al. 2004), HASM3 (Yue and Du 2005), and HASM4 (Yue et al. 2007a). HASM1 separately focuses on each of the Gauss equations to determine the method with the highest accuracy, and HASM1a, HASM1b, and HASM1c exist. However, its accuracy is not ideal. HASM2 considers all the three Gauss equations, and the error is remarkably lower than that of HASM1, but calculation overflow occurs in the simulation. HASM3 is based on the first two Gauss equations and has a higher simulation accuracy than HASM2. However, since the inverse of the matrix is recalculated during each iteration, the computational burden is high. HASM4, improved from HASM3, has a higher accuracy and a lower computational burden. The accuracy of HASM improves gradually from stage 1 to stage 4; in particular, the computational speed of HASM4 is much faster than those of HASM1, HASM2, and HASM3 (Yue and Du 2006a). Numerical results show that the accuracy of HASM is several orders of magnitude higher than those of the classical interpolation methods widely used in GIS and computer-aided design (CAD), e.g., the inverse distance weight (IDW) method, the kriging method, and the spline method (Yue et al. 2007). According to a theoretical analysis, HASM avoids the peak-cutting issue in numerical simulation, and the simulation accuracy is not very sensitive to the distance between sampling points (Yue and Du 2005). The entire computational process of HASM can be divided into three stages, i.e., the discretization of partial differential equations, the establishment of sampling equations, and the solving of algebraic equations (Yue 2011). In other words, based on the principal theorem of surface theory, HASM discretizes the differential equations satisfied by the surface and then solves the discrete algebraic system per the information about the sampling points.

Traditional HASM

According to differential geometry, a spatial surface is determined by the first and second fundamental coefficients. If the surface is expressed as $z = f(x, y)$, the basic theory of traditional HASM can be represented as follows (Yue et al. 2004, 2007; Yue and Du 2005, 2006a, b; Yue 2011):

$$\begin{cases} f_{xx} = \Gamma_{11}^1 f_x + \Gamma_{11}^2 f_y + \frac{L}{\sqrt{E+G-1}} \\ f_{yy} = \Gamma_{22}^1 f_x + \Gamma_{22}^2 f_y + \frac{N}{\sqrt{E+G-1}} \end{cases} \tag{1.5}$$

where

$$E = 1 + f_x^2, \, F = f_x \cdot f_y, \, G = 1 + f_y^2, \, L = \frac{f_{xx}}{\sqrt{1 + f_x^2 + f_y^2}}, \, N = \frac{f_{yy}}{\sqrt{1 + f_x^2 + f_y^2}},$$

$$\Gamma_{11}^1 = \frac{1}{2}\left(GE_x - 2FF_x + FE_y\right)\left(EG - F^2\right)^{-1},$$

$$\Gamma_{11}^2 = \frac{1}{2}\left(EF_x - EE_y - FE_x\right)\left(EG - F^2\right)^{-1},$$

$$\Gamma_{22}^1 = \frac{1}{2}\left(2GF_y - GG_x - FG_y\right)\left(EG - F^2\right)^{-1},$$

$$\Gamma_{22}^2 = \frac{1}{2}\left(EG_y - 2FF_y + FG_x\right)\left(EG - F^2\right)^{-1}.$$

For grids with a spatial step size of h, $\{(x_i, y_j)|0 \le i \le I + 1, 0 \le j \le J + 1\}$. The discretization of the first-order partial derivatives f_x, f_y and the second-order partial derivatives f_{xx}, f_{yy} is expressed as follows:

$$(f_x)_{(i,j)} \approx \begin{cases} \frac{f_{1,j} - f_{0,j}}{h} i = 0 \\ \frac{f_{i+1,j} - f_{i-1,j}}{2h} i = 1, \cdots, I, \\ \frac{f_{I+1,j} - f_{I,j}}{h} i = I + 1 \end{cases} \quad (f_{xx})_{(i,j)} \approx$$

$$(f_x)_{(i,j)} \approx \begin{cases} \frac{f_{0,j}+f_{2,j}-2f_{1,j}}{h^2} & i = 0 \\ \frac{f_{i-1,j}-2f_{i,j}+f_{i+1,j}}{2h^2} & i = 1, \cdots, I \\ \frac{f_{I+1,j}+f_{I-1,j}-2f_{I,j}}{h^2} & i = I+1 \end{cases}$$

$$(f_y)_{(i,j)} \approx \begin{cases} \frac{f_{i,1}-f_{i,0}}{h} & j = 0 \\ \frac{f_{i,j+1}-f_{i,j-1}}{2h} & j = 1, \cdots, J, \\ \frac{f_{i,J+1}-f_{i,J}}{h} & j = J+1 \end{cases} \quad (f_{yy})_{(i,j)} \approx \begin{cases} \frac{f_{i,0}+f_{i,2}-2f_{i,1}}{h^2} & j = 0 \\ \frac{f_{i,j-1}-2f_{i,j}+f_{i,j+1}}{2h^2} & j = 1, \cdots, J. \\ \frac{f_{i,J+1}+f_{i,J-1}-2f_{i,J}}{h^2} & j = J+1 \end{cases}$$

$$(1.6)$$

The iterative differential equations of the partial differential equations corresponding to (1.5) are as follows:

$$\begin{cases} \dfrac{f_{i+1,j}^{n+1} - 2f_{i,j}^{n+1} + f_{i-1,j}^{n+1}}{h^2} = (\Gamma_{11}^1)_{i,j}^n \dfrac{f_{i+1,j}^n - f_{i-1,j}^n}{2h} + (\Gamma_{11}^2)_{i,j}^n \dfrac{f_{i,j+1}^n - f_{i,j-1}^n}{2h} + \dfrac{L_{ij}^n}{\sqrt{E_{i,j}^n + G_{i,j}^n - 1}} \\[4mm] \dfrac{f_{i,j+1}^{n+1} - 2f_{i,j}^{n+1} + f_{i,j-1}^{n+1}}{h^2} = (\Gamma_{22}^1)_{i,j}^n \dfrac{f_{i+1,j}^n - f_{i-1,j}^n}{2h} + (\Gamma_{22}^2)_{i,j}^n \dfrac{f_{i,j+1}^n - f_{i,j-1}^n}{2h} + \dfrac{N_{ij}^n}{\sqrt{E_{i,j}^n + G_{i,j}^n - 1}} \end{cases}$$

$$(1.7)$$

$$E_{i,j}^n = 1 + \left(\frac{f_{i+1,j}^n - f_{i-1,j}^n}{2h} \right)^2, \quad F_{i,j}^n = \left(\frac{f_{i+1,j}^n - f_{i-1,j}^n}{2h} \right) \left(\frac{f_{i,j+1}^n - f_{i,j-1}^n}{2h} \right),$$

$$G_{i,j}^n = 1 + \left(\frac{f_{i,j+1}^n - f_{i,j-1}^n}{2h} \right)^2,$$

$$L_{i,j}^n = \frac{\dfrac{f_{i-1,\,j}^n - 2f_{i,\,j}^n + f_{i+1,\,j}^n}{2h^2}}{\sqrt{1 + \left(\dfrac{f_{i+1,\,j}^n - f_{i-1,\,j}^n}{2h} \right)^2 + \left(\dfrac{f_{i,j+1}^n - f_{i,\,j-1}^n}{2h} \right)^2}},$$

$$N_{i,j}^n = \frac{\dfrac{f_{i,\,j-1}^n - 2f_{i,\,j}^n + f_{i,\,j-1}^n}{2h^2}}{\sqrt{1 + \left(\dfrac{f_{i+1,\,j}^n - f_{i-1,\,j}^n}{2h} \right)^2 + \left(\dfrac{f_{i,j+1}^n - f_{i,\,j-1}^n}{2h} \right)^2}},$$

$$(\Gamma_{11}^1)_{i,j}^n = \frac{G_{i,j}^n (E_{i+1,j}^n - E_{i-1,j}^n) - 2F_{i,j}^n \left(F_{i+1,j}^n - F_{i-1,j}^n \right) + F_{i,j}^n \left(E_{i,j+1}^n - E_{i,j-1}^n \right)}{4h \left(E_{i,j}^n G_{i,j}^n - (F_{i,j}^n)^2 \right)},$$

$$(\Gamma_{11}^2)_{i,j}^n = \frac{2E_{i,j}^n \left(F_{i+1,j}^n - F_{i-1,j}^n \right) - E_{i,j}^n (E_{i,j+1}^n - E_{i,j-1}^n) - F_{i,j}^n \left(E_{i,j+1}^n - E_{i,j-1}^n \right)}{4h \left(E_{i,j}^n G_{i,j}^n - (F_{i,j}^n)^2 \right)},$$

$$(\Gamma_{22}^1)_{i,j}^n = \frac{2G_{i,j}^n \left(F_{i,j+1}^n - F_{i,j-1}^n \right) - G_{i,j}^n (G_{i+1,j}^n - G_{i-1,j}^n) - F_{i,j}^n \left(G_{i,j+1}^n - G_{i,j-1}^n \right)}{4h \left(E_{i,j}^n G_{i,j}^n - (F_{i,j}^n)^2 \right)},$$

$$\left(\Gamma^2_{22}\right)^n_{i,j} = \frac{E^n_{i,j}\left(G^n_{i,j+1} - G^n_{i,j-1}\right) - 2F^n_{i,j}(F^n_{i,j+1} - F^n_{i,j-1}) + F^n_{i,j}\left(G^n_{i+1,j} - G^n_{i-1,j}\right)}{4h\left(E^n_{i,j}G^n_{i,j} - (F^n_{i,j})^2\right)}.$$

$f^{n+1}_{0,j} = f^0_{0,j}(0 \leq j \leq J + 1);\ f^{n+1}_{i,0} = f^0_{i,0}(0 \leq i \leq I + 1);\ f^{n+1}_{I+1,j} = f^0_{I+1,j}(0 < j < J + 1);\ f^{n+1}_{0,J+1} = f^0_{i,J+1}(0 < i < I + 1);$
$f^{n+1}_{0,j},\ f^{n+1}_{i,0},\ f^{n+1}_{I+1,j},$ and $f^{n+1}_{i,J+1}$ are the boundary conditions of HASM.

The matrix expressions of (1.7) are as follows:

$$\begin{cases} Az^{n+1} = \mathbf{d}^n \\ Bz^{n+1} = \mathbf{q}^n \end{cases} \tag{1.8}$$

where

$$A = \begin{bmatrix} -2I_{J\times J} & I_{J\times J} & & & \\ I_{J\times J} & -2I_{J\times J} & I_{J\times J} & & \\ & O & O & O & \\ & & I_{J\times J} & -2I_{J\times J} & I_{J\times J} \\ & & & I_{J\times J} & -2I_{J\times J} \end{bmatrix},$$

$$B \begin{bmatrix} B_{J\times J} & & \\ & B_{J\times J} & \\ & & B_{J\times J} \end{bmatrix}, B_{J\times J} \begin{bmatrix} -2 & 1 & & \\ 1 & -2 & 1 & \\ O & O & O & \\ & & & -2 & 1 \end{bmatrix}$$

$$z^{n+1} = \left(f^{n+1}_{1,1}, \cdots, f^{n+1}_{1,J}, f^{n+1}_{2,1}, \cdots, f^{n+1}_{2,J}, \cdots\cdots\cdots, \right.$$
$$\left. f^{n+1}_{I-1,1}, \cdots, f^{n+1}_{I,1}, \cdots, f^{n+1}_{I,J}\right)^T.$$

A and $\mathbf{d}^n$ are the matrices of the coefficient and the right constant term of the first equation in (1.8), respectively, and **B** and $\mathbf{q}^n$ are the matrices of the coefficient and the right constant term of the second equation in (1.8), respectively.

To ensure that the true value of the sampling point is equal to or close to the estimated value of the sampling point, the expression of HASM is transformed into a least square problem with equality constraints.

$$\begin{cases} \min \left\| \begin{bmatrix} A \\ B \end{bmatrix} z^{n+1} - \begin{bmatrix} d \\ q \end{bmatrix}^n \right\|_2 \\ \text{s.t.} Sz^{n+1} = k \end{cases} \tag{1.9}$$

where **S** and **k** represent the coefficient matrix and value of the sampling point and can be expressed as $S(t, (i - 1) \cdot J + j) = 1$ and $\mathbf{k}(t) = \bar{f}_{i,j}$, respectively; that is, the value of the t-th sampling point (x_i, y_j) is $\bar{f}_{i,j}$.

Fig. 1.1 Distribution of nonzero elements in the HASM matrix

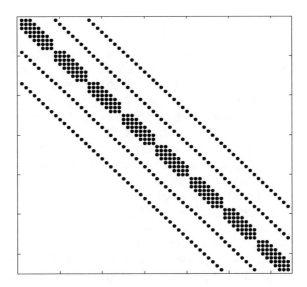

To solve the constrained least square problem in (1.9), a positive parameter λ, which is the weight of the sampling point and is determined by the contribution of the sampling point to the simulated surface, is introduced (Yue 2011). For a given λ, (1.9) can be transformed into an unconstrained least square problem.

$$\min \left\| \begin{bmatrix} \mathbf{A} \\ \mathbf{B} \\ \lambda\mathbf{S} \end{bmatrix} z^{n+1} - \begin{bmatrix} \mathbf{d}^n \\ \mathbf{q}^n \\ \lambda\mathbf{k} \end{bmatrix} \right\|_2 \tag{1.10}$$

This is then further transformed into the following equation.

$$\mathbf{W}z^{n+1} = \mathbf{v}^n \tag{1.11}$$

where $\mathbf{W} = \begin{bmatrix} \mathbf{A}^T & \mathbf{B}^T & \lambda\mathbf{S}^T \end{bmatrix} \begin{bmatrix} \mathbf{A} \\ \mathbf{B} \\ \lambda\mathbf{S} \end{bmatrix}$, $\mathbf{v}^n = \begin{bmatrix} \mathbf{A}^T & \mathbf{B}^T & \lambda\mathbf{S}^T \end{bmatrix} \begin{bmatrix} \mathbf{d}^n \\ \mathbf{q}^n \\ \lambda\mathbf{k} \end{bmatrix}$. Figure 1.1 shows the distribution diagram of nonzero elements of $\mathbf{W}$ ($I = J = 8$, $I \times J = 64$, the order of $\mathbf{W}$ is 64×64), and nz denotes the number of nonzero elements.

Based on the sampling points, traditional HASM simulates the spatial distribution of a surface through a series of iterations and then interpolates the values at unsampled points.

1.2.2.3   Theoretical Problems Faced by Traditional HASM

Traditional HASM has some shortcomings, such as massive storage requirements and a high computational burden. To increase the computational speed of HASM and improve its data processing ability, many improvements have been proposed in terms of different aspects of the HASM process, such as the multigrid algorithm, adaptive algorithm, improved Gauss–Seidel algorithm and diagonal-based preconditioned conjugate gradient (CG) method (Yue and Song 2008; Song and Yue 2009; Yue and Wang 2010; Yue and Wang et al. 2010; Yue et al. 2012). Based on the variations in spatial grid resolution, the multigrid method and the adaptive algorithm continuously adjust the entire simulation process according to the actual needs, and the direct method or the iterative method is used to solve the algebraic equations on each layer of grids. Under the fixed grid resolution, the improved Gauss–Seidel algorithm and the preconditioned CG algorithm can accelerate the convergence rate of the large sparse algebraic equations formed in HASM. These methods have greatly improved the iterative speed to varying degrees. The performance of different methods greatly differs, and different methods are applicable to different scenarios.

In the multigrid algorithm, the low-frequency and high-frequency components of the interpolation error are relative, and both are related to the grid size. A smooth component on fine grids may be regarded as an oscillating component on coarse grids. In the multigrid method, the smooth components on the fine grids can be eliminated on the coarse grids using an iterative method layer by layer until all the error components are eliminated. Thus, the multigrid method can effectively eliminate high-frequency errors and is more suitable for studying the spatial interpolation of meteorological elements on a large time scale (Yue et al. 2013a, b). However, under random sampling conditions, when the sampling point is not in the centre of the pixel, the coarser the grid is, the greater the error caused by the spatial location. In addition, when the component is being transferred from the coarse grid to the fine grid, the error could be transferred to the fine grid as well, lowering the simulation accuracy. To reduce the influence of spatial location error, the grid layers are reduced, but the simulation time increases accordingly, which weakens the advantages of the method.

The adaptive method selects different grid resolutions according to the terrain complexity of the simulated area; fine grids are used in areas with complicated terrain, and coarse grids are used in flat areas. The reduction in the computation time with the coarse grids leads to a decrease in the computation time for the adaptive method. However, the setting of fine and coarse grids is rather subjective, and different grid resolutions have a strong influence on the simulation error. Since the grid variation in the adaptive method should be based on some physical properties of the solution, this method needs to repeatedly test the grid variation based on different practical problems. When the adaptive method is applied to HASM, the selection of the initial grid resolution greatly influences the final results, and a numerical solution on the initial grid is required to capture the basic information of the original data to the greatest extent. Moreover, during grid refinement, the inaccurate determination of the grid areas that need to be further refined can lead to the failure of the following work.

Both the multigrid method and the adaptive method need to use different iterative algorithms, such as the Gauss–Seidel algorithm, to solve the corresponding equations. The improved Gauss–Seidel algorithm directly accelerates the solving of algebraic equations on the grid with the same step size, which simplifies the HASM process. Hence, solving the HASM equations is the key to improving the convergence speed of HASM. By comparing the improved Gauss–Seidel method with the preconditioned CG method, Chen (2010) concluded that the convergence of HASM by the diagonal preconditioned CG method was faster than that of HASM by the Gauss–Seidel algorithm. The doctoral dissertation of Yan (2012) on the computational performance of the improved Gauss–Seidel algorithm and the preconditioned CG method pointed out that the preconditioned CG method converged faster than the improved Gauss–Seidel algorithm. Numerical simulations by Chen and Yue (2010a, b) showed that the preconditioned CG method was much faster than the classical iterative method in MATLAB in solving the HASM equations. Although the preconditioned CG method based on diagonals increases the solving speed of HASM to some extent, previous studies were only based on simulations; the root cause of the improvement in the computational speed of HASM was not analysed from the perspective of the HASM equations themselves. In addition, studies have shown that the computational speed of HASM can be further improved.

Although the simulation accuracy of traditional HASM is much higher than those of classical interpolation methods, there are several theoretical problems in the application of traditional HASM.

1. Although HASM is based on the principal theorem of surface theory, its theoretical basis is incomplete. During the early development stage of HASM, all partial differential equations of the surface are considered, but data overflow occurs during computation. To prevent this, the traditional HASM considers only the first two partial differential equations of the surface (Yue and Du 2006a), so traditional HASM is only based on a part of the principal theorem of surface theory. There are three Gauss equations, so the traditional HASM fails to fully describe the content expressed by the principal theorem of surface theory, which leads to unstable performance in practical applications.

2. An oscillating boundary issue occurs in the simulated area, and HASM does not fully solve this boundary error, so the simulation accuracy at the boundary remains low.

3. Currently, the driving field of HASM is calculated using other interpolation methods based on the sampling data. For a randomly selected driving field, the simulation accuracy of HASM is not ideal, and its performance is often worse than that of the other interpolation methods. Traditional HASM is sensitive to the driving field, and the iterative driving field is provided by other interpolation methods. Consequently, traditional HASM is often regarded as a modification of interpolation methods, which limits its popularization and application.

4. HASM is divided into two parts, i.e., the inner iteration and outer iteration. The outer iteration mainly modifies the right end term of the HASM equations so that the first and second fundamental coefficients of the surface satisfy the

Gauss-Codazzi equation. The inner iteration refers to the process of solving the HASM equations. The stopping criteria for the inner iteration of HASM are currently based on the iteration convergence criteria of the equations, while the stopping criteria for the outer iteration lack a relevant basis, and a certain number of iterations is often set by programmers according to their experience, which lacks a theoretical basis. It is difficult to set the number of outer iterations for different problems, and users often find it hard to judge whether the final surface obtained by HASM is the surface that they need.

1.2.3   Main Spatial Interpolation Methods of Climatic Elements

In reality, for many existing problems, data interpolation is often the first thing that needs to be achieved before analysis and research based on the complete data. Spatial interpolation uses the values of meteorological elements collected at meteorological stations to obtain the values of meteorological elements in areas without meteorological stations. This is the primary problem in the study of the processing of land surfaces. Studies have shown that there is no optimal spatial interpolation method for meteorological elements for all cases.

The simplest spatial interpolation method of meteorological elements is the assignment of the value of a surrounding known point to the point to be estimated. The use of the Thiessen polygon (Tiessen 1911; Goovaerts 1999b), which is also known as the nearest neighbour method (Hartkamp et al. 1999), is one such method. During interpolation, the variable value of the point to be estimated should be the same as that of the nearest observation point. Creutin and Obled (1982), Tabios and Salas (1985), McCuen (1998), and Dirks et al. (1998) applied this method to precipitation estimation. In relatively small areas where the spatial variability in variables is insignificant, the interpolation result is acceptable and easy to achieve. However, the spatial distribution of precipitation is often very uncertain and spatially non-stationary, and this method could yield very large errors.

The IDW algorithm is another simple and feasible interpolation method (Wei and McGuinness 1973; Simanton and Osborn 1980; Creutin and Obled 1982; Dodson and Marks 1997; Bedient and Huber 1992; Lu and Wong 2008; Singh 2019; Gu 2020; Solmaz 2020; Bărbulescu 2021; Masoudi 2021). Dale et al. (1999) verified the effectiveness of the algorithm based on synthetic data, and the results showed that inverse distance interpolation using 6 adjacent points outperforms that using 12 adjacent points. Caruso and Quarta (1998) concluded that the interpolation of the inverse distance function is closely related to the selected parameters. If the observation points are dense within the simulated area, good results can be achieved by selecting a relatively small radius for computation. Richard (1982) proposed that to calculate the weight, the same order can be used for all distances, or different orders can be used for different distances. IDW is simple and practical in the exploratory analysis of spatially continuous data. As an interpolation method at a global scope, IDW can be regionalized using different methods. Nevertheless, IDW has many limitations. For example, the number of sample points in the local neighbourhood needed to estimate an unknown data point remains unknown; when there is spatial heterogeneity or anisotropy in the element field, the estimation can be affected by the size, direction, and shape of the neighbourhood; the estimation of the weight coefficient depends on experience, which has no sufficient theoretical basis; the estimated data of the unknown points cannot exceed the range of the observed values; and the results are determined by the layout of the sampling points. The IDW method fails to consider the relationship between spatial structure information and information beyond the neighbourhood. Through further studies on the IDW, Teegavarapu and Chandramouli (2005) and Tomczak (1998) proposed some improvement strategies. Willmott and Robseon (1995) proposed a temperature-assisted IDW interpolation method based on high-resolution temperature data from meteorological stations and an annual average temperature deviation gained by spatial interpolation. When studying the interpolation on climate data, in addition to the distance weight, Nalder and Wein (1998) considered the gradient changes in meteorological elements with altitude, longitude, and latitude and proposed the gradient inverse distance square (GIDS) method. Lu and Wong (2008) improved the weight coefficient in IDW by introducing a delay parameter for the reduction in the degree of spatial autocorrelation and proposed an adaptive IDW algorithm. Bărbulescu (2021) computed the Beta Parameter in IDW Interpolation by Using a Genetic Algorithm.

The kriging method (Chiles 1999) has been extensively used because it considers the spatial correlation between spatial variables. The kriging method and its various forms are widely used in precipitation interpolation (Matheron 1971; Krajewski 1987; Dingman et al. 1988; Hevesi et al. 1992; Phillips et al. 1992; Seo and Smith 1993; Garen et al. 1994; Seo 1996; Ashraf et al. 1997; Goovaerts 1999b; Vieux 2001; Grayson and Bloschl 2001; Dingman 2002; Teegavarapu 2007; Hofstra et al. 2008; Sun et al. 2010; Feng 2019; Hou 2020; Wang 2020; Xiong 2021). Theoretically, the kriging method realizes the best linear unbiased optimal estimation; however, it also has limitations. First, the kriging method is based on the second-order stationary hypothesis of spatial variables; second, it requires a single semi-variance function to describe the spatial variations in data attributes. However, it is basically impossible to satisfy these two requirements in practice. Moreover, the kriging method assumes a linear relationship between the explanatory variables and the predicted variables, which is often not the case in reality (Leathwick et al. 2006; Hjort and Luoto 2010). The co-kriging method not only considers the spatial correlation of variables but also makes full use of other auxiliary information, which improves the interpolation accuracy. Yang et al. (2009), and Jiang et al. (2010) applied the co-kriging method to the interpolation of temperature data and obtained great results. Aalto et al. (2012) used the kriging with external drift to predict the monthly average precipitation in Finland and obtained promising results. Rizzo and Dougherty (1994), Lin and Chen (2004), and Koike et al. (2001) combined a neural network with the kriging method to study the spatial distribution of meteorological elements. When interpolating the monthly average temperature in Heilongjiang Province, Li et al. (2012) extended the kriging method from the spatial dimension to the temporal dimension and pointed out that the improved kriging method outperforms the spatial kriging method. By incorporating the parameter-elevation regression on independent slopes model (PRISM) into kriging interpolation, Wang et al. (2011a, b) proposed the precipitation-elevation regressions of kriging (PER-kriging) interpolation method, which can effectively eliminate the abnormal interpolation results caused by the very large terrain differences between the observation points and the surrounding area. Bai et al. (2011) parallelized the kriging method using a message passing interface (MPI) to reduce the computation time of the kriging method in spatial precipitation interpolation. Wang (2020) improved RSS Data Generation Method Based on Kriging Interpolation Algorithm.

Due to the resulting smoothness, the spline function method has been used in many studies on the interpolation of meteorological elements (Hutchinson and Gessler 1994; Hutchinson 1991, 1995; Hulme et al. 1995; Lennon and Turner 1995; Hutchinson 1998a,b; Jeffrey et al. 2001; Jarvis and Stuart 2001a, 2001b; Yan et al. 2005; New et al. 2002; Hijmans et al. 2006; Zhu et al. 2005a, b; Li et al. 2007; Ma et al. 2008; Zhu et al. 2009; Qian et al. 2010; Kawser 2019; Deepti 2020; Roe 2021). Hutchinson (1991) believed that the local thin plate smoothing spline method has promising application prospects when data reliability and density are limited. When analysing precipitation interpolation with the thin plate smoothing spline method, Hutchinson (1995) proposed a spline model under different parameter settings and pointed out that the spline model considering only the influence of elevation on precipitation is suitable for the estimation of monthly precipitation. Jeffrey et al. (2001) applied the thin plate smoothing spline method to interpolate Australian daily climate data. Hutchinson (2001) used the software Aunsplin to interpolate precipitation data with longitude, latitude, and elevation as independent variables. Yan (2004) and Yan et al. (2005) simulated the spatial distribution of climatic elements in China using Aunsplin and verified the applicability of this method. Hancock and Hutchinson (2006) developed the thin plate smoothing spline method and used it to process large-scale meteorological interpolation. Liu et al. (2008) pointed out that Aunsplin is more suitable for the interpolation of meteorological elements of time series. Liu et al. (2012) used Aunsplin to interpolate the annual average precipitation in China and analysed the influences of the elevation and the distance from the coastline on the interpolation results. Venier et al. (1998a, b) used Aunsplin to interpolate the climatic elements in Lake Ontario to predict the influence of climate change on species distributions. However, Aunsplin requires constant manual intervention during simulation to eliminate the stations with extensive noise, and the optimal fitting effect of a continuous surface is achieved by minimizing the value of the generalized cross-validation. As pointed out by Collins and Bolstad (1996) and Hartkamp et al. (1999), the spline method is an ideal method when meteorological stations are regularly and densely distributed.

Although the above classical interpolation methods have been extensively applied to the interpolation of meteorological elements, these methods are based on spatial autocorrelation or continuous spatial smoothing without considering the influence of relevant factors. The interpolation accuracy is low, and the physical significance of the model is not clear. Due to the limited number of meteorological stations and the complexity and diversity of factors affecting the spatial distribution of temperature and precipitation, it is difficult to accurately obtain the spatial distribution of actual temperature and precipitation in a specific area. With the development of GIS and computer technologies as well as widespread attention from the academia worldwide, many interpolation models and improved algorithms have been developed. Interpolation simply based on meteorological data could inevitably result in interpolation errors. Multivariate regression is an interpolation method that describes the relationship between related variables using mathematical expressions. Many studies have been conducted on the relationships between temperature, precipitation, and their influencing factors using multivariate regression. A regression model can be established based on the geographical coordinates and elevation information of meteorological stations as well as the other influencing factors, such as aspect and slope. Meteorological elements change with elevation. Generally, the temperature decreases with increasing elevation, while precipitation is positively correlated with elevation and starts to decrease with elevation only after a certain height is reached (Schermerhorn 1967; Dingman 1981; Daly et al. 1994; Fu 1983; Tang 1985; Wang and Cheng 1998). The distribution pattern of precipitation in mountainous areas varies in a more complex way with elevation, and there are great differences in different regions (Lin 1995). Many studies have shown that errors in digital elevation model (DEM) data have an impact on rasterization results in temperature interpolation (Houghton 1979; Smith 1979; Hevesi et al. 1992; Daly et al. 1994; Martinez-Cob 1996; Prudhomme and Duncan 1999; Kurtzman and Kadmon 1999; Goovaert 2000; Ninyerola et al. 2000; Weisse and Bois 2001; Hofierka et al. 2002; Diodato and Ceccarelli 2005; Apaydin et al. 2011; Liao and Li, 2004; You and Li 2005; Lioyd 2005). In some studies, the influences of topographic factors, such as latitude, elevation, continentality, slope, and topographic openness, and land landscape factors are considered during the interpolation of meteorological data (Basist et al. 1994; Ollinger et al. 1993, 1995; Benzi et al. 1997; Chessa and Delitala 1997; Vogt et al. 1997; Goodale et al. 1998; Ninyerola et al. 2000; Wotling et al. 2000; Weisse and Bois 2001; Oettli and Camberlin 2005; Yue et al. 2005; Zeng and Huang 2007; Apaydin et al. 2011; Hao et al. 2020; Uribe 2021; Raveesh et al. 2021). Based on Landsat satellite remote sensing data and data collected at meteorological stations, Eleanor et al. (1984) used the multivariate regression method to explore the spatial distribution of temperature and precipitation with consideration of the impact of topographic variables on temperature and precipitation. Burrough (1996) pointed out that the method could be accurate enough as long as sufficient terms and factors are considered in polynomial regression; however, the physical meaning of the model is often unclear (Lin et al. 2002). Goodale et al. (1998) simulated temperature and precipitation with respect to longitude, latitude and elevation using quadratic polynomial regression, and the results showed that this method consumes less time and memory. Yan et al. (2005) reported that introducing more explanatory variables into the regression can lead to a rapid increase in computational cost. Marquinez et al. (2003) analysed the relationships between precipitation and topographic variables by multivariate regression and GIS technology and concluded that the optimal interpolation model should consider elevation, slope, aspect, distance from the coastline, and distance from the west as influencing factors. According to Goovaerts (2000), Price et al. (2000), and Yan et al. (2005), longitude, latitude, and elevation are the main explanatory variables in the spatial interpolation of monthly average temperature and precipitation, and other variables, such as continentality, also play a critical role in interpolation (Holdaway 1996; Vajda and Venalainen 2003). A study conducted by Attorre et al. (2007) showed that an interpolation method considering meteorological factors is more accurate than an interpolation method only using data collected by meteorological stations.

Daly et al. (1994) established the PRISM model based on multivariate regression. The model, as a multivariate regression method, can interpolate temperature and precipitation on different time scales. In this model, elevation is considered the primary factor influencing precipitation; a special meteorological element-elevation linear regression function is calculated for each grid. PRISM is suitable for areas with topographic inequalities (Schwarb 2001; Szentimrey et al. 2007). The model (Szentimrey et al. 2007) has been widely used in climate-related mapping, with the main achievements including the 103-year time series dataset of the climate in the United States and its neighbouring regions (Daly et al. 2000a, b) and the climate atlas newly published by the National Climatic Data Center of the United States (Plantico et al. 2000; Daly et al. 2000a, b). Zhu et al. (2003) used PRISM to simulate the temperature and precipitation in China, and the results showed that PRISM can effectively simulate the spatial distribution and seasonal variation in temperature and precipitation in China, except for mountainous and subtropical areas, where the simulation results might by affected by the differences in surface cover and topography. Zhao et al. (2004) and Zhu et al. (2005a, b) used PRISM to study the spatial interpolation of daily precipitation in the middle part of Heihe River Basin at Hexi Corridor, and the results implied that this simple and reliable method can meet the requirements of distributed hydrological models or related land surface process simulations regarding the spatiotemporal accuracy of daily precipitation.

The spatial variation surfaces of any factor in nature can be divided into two types, namely, drastic variation and mild variation. Any spatial variation can be regarded as a combination of these two types, with different ratios of these two. The global interpolation method can be used for the mild variation part (the part with low-frequency oscillation), and the local interpolation method can be applied to the part with drastic variation (the part with high-frequency oscillation). Hybrid interpolation is a widely used interpolation method that combines the strengths of global interpolation and local interpolation. As pointed out by Shi (2010), hybrid interpolation can be decomposed into the sum of different weights of global interpolation and local interpolation, while the selection of weights is determined by the part with low-frequency oscillation. Hence, the accuracy of the trend surface is very significant to the final interpolation results in hybrid interpolation. Bloomfield (1992), Bloomfield and Nychka (1992), Woodward and Gray (1993), Willmott and Matsuura (1995), Holdaway (1996), and Perry (2006) all used hybrid interpolation to simulate the spatial distribution of meteorological elements. Burrough and McDonnell (1998) showed that hybrid interpolation not only generates a continuously changing surface of meteorological elements but also provides the relationships between meteorological elements and their influencing factors. Ninyerola et al. (2007) pointed out that residual correction after removing the trend surface not only reduces the error in trend surface analysis but also improves the final simulation accuracy. By comparing the regression method and the kriging method, Joly et al. (2011) presented the shortcomings of the two methods and suggested that hybrid interpolation should be applied to make up for the limitations. Liu et al. (2004) used a hybrid interpolation method combining three-dimensional (3D) quadratic trend surface analysis and IDW to simulate the spatial distribution of temperature and precipitation in China. He et al. (2005) pointed out that the hybrid interpolation method improves the accuracy of precipitation simulation and can be an overall direction for future studies on precipitation interpolation. However, the trend surface analysis in hybrid interpolation is often based on the principle of least squares. The classical linear regression model requires the residual to have no autocorrelation; however, the residual after trend surface analysis often features autocorrelation to varying degrees. Anselin and Rey (1991) pointed out that once the residual has spatial autocorrelation, the classical linear regression estimation is no longer accurate. To express the spatial differences among research objects, Brunsdon et al. (1996) and Fotheringham (2000) developed geographically weighted regression analysis by adding a relational matrix based on spatial location to the classical regression model, which allows the existence of different spatial relationships in different locations; therefore, the analysis results are regional, and the spatial non-stationarity of the data can be detected by this model. As the core of this method, the spatial weight matrix is determined by the spatial location and direction of the research object. If the diagonal element is 1, the spatial object has global consistency and no spatial variation and correlation, and the geographically weighted regression model is degenerated into classical regression analysis at this time. Zhang et al. (2005) mentioned that geographically weighted regression can effectively reduce the spatial autocorrelation of residuals. Through geographically weighted regression, Brunsdon et al. (2001) analysed the relationship between the total annual precipitation and the station elevation in the UK and studied the variations in precipitation with elevation.

To achieve high interpolation accuracy in the interpolation of meteorological elements, methodological studies must be carried out, especially on the relationships between meteorological elements and their influencing factors. The spatial interpolation of meteorological elements tends to be based on multivariable analysis and complex computation. Dietz et al. (2007) explored the influence of human activities on climate. Mebrhatu et al. (2004) studied the effect of sea surface temperature on precipitation. Solantie (1976), Felicisimo Perez (1992), Holdaway (1996), Fernandez Alvarez (1996), Griffiths and McSaveney (1983), and Marquinez et al. (2003) pointed out that the distance from the water body greatly influences the distribution of precipitation. Zheng et al. (1997) and Zheng and Basher (1998) explored the influence of the southern oscillation index (SOI) on precipitation interpolation. In a study on the interpolation of annual precipitation in Xinjiang, China, Wang et al. (2011a, b) studied the influence of the normalized difference vegetation index (NDVI) on precipitation. When simulating the distribution of temperature in mountainous areas, Chen et al. (1998) investigated the effect of direct solar radiation on temperature by multivariate regression. In addition, some new methods have been developed. French et al.

(1992), Govindaraju and Rao (2000), Teegavarapu and Chandramouli (2005), Mo and Zhang (2007), Wang et al. (2007), and Zhang and Liao (2011) applied artificial neural networks to the spatial interpolation of precipitation, and the results showed that artificial neural networks can improve the accuracy of precipitation interpolation. Lin and Chen (2004) proposed the improved radial basis function network (IRBFN) for interpolation, which combines the standard radial basis function network with the semi-variance function model. The IRBFN interpolation method was used to analyse the spatial distribution of precipitation in the Tanshui River area in South Taiwan. The comparisons between the IRBFN interpolation method, standard radial basis function network, and ordinary kriging interpolation method showed that the IRBFN interpolation method has the smallest interpolation error. In a study on the annual precipitation in Xinjiang, China, Wang et al. (2011a, b) applied multivariate regression to increase the number of simulated meteorological stations and used the multiquadric radial basis function (MRBF) to interpolate the precipitation, yielding good interpolation results. By analysing the inherent relationships between the spatial distribution of the monthly and annual average temperature in China and the longitude, latitude, and elevation and the optimal response distance among meteorological stations during interpolation, Pan et al. (2004) proposed a smart spatial interpolation (SSI) method for temperature based on DEM and smart search distance and compared this method with traditional methods such as IDW. The interpolation results obtained by the proposed SSI method not only are highly accurate but also reveal the zonal characteristics of the temperature changes with latitude, longitude, and altitude in detail. By setting up virtual stations, Wang et al. (2011a, b), Lu et al. (2010), Xu et al. (2008), and Gu et al. (2006) studied the spatial interpolation of meteorological elements, especially those in sparse areas, and the results suggested that the density of meteorological stations has a great influence on the interpolation accuracy.

    The spatial distribution of meteorological elements in different regions is influenced by different factors, and the spatial variation and characteristics of different regionalized factors are different. Therefore, the spatial interpolation methods and models suitable for different regions are also different. Different methods approximate the distribution characteristics of interpolation elements in different ways. To select the optimal interpolation model, the characteristics of different interpolation models and the distribution density and regional characteristics of the stations should be understood. The key to the spatial interpolation of meteorological elements is still to select an interpolation method suitable for the spatial distribution and variation of all meteorological elements. The selection of the interpolation method depends on the type of data used, the study area, the temporal and spatial scale, and the spatial distribution pattern of regionalized variables (Brus and Heuvelink 2007). Different methods yield different interpolation results for the same area, and no interpolation method can remain optimal for all cases (Lam 1983; Bastin et al. 1984; Tabios and Salas 1985; Hevesi et al. 1992. Hutchinson 1995, 1998a, b; Hay et al. 1998; Goovaerts 2000; Deraisme et al. 2001; Hofierka et al. 2002; Attorre et al. 2007; Ninyerola et al. 2007; Portales et al. 2010; Lin et al. 2002; He et al. 2005). The key to the interpolation of meteorological elements is to select the optimal method based on specific data and

background (Burrough and McDonnell 1998). By studying precipitation interpolation, Creutin and Obled (1982), Tabios and Salas (1985), Phillips et al. (1992) found
that the kriging method outperforms the IDW method and the Thiessen polygon
method. Tobin et al. (2011) interpolated the temperature and precipitation in mountainous areas in Switzerland using the IDW method, the kriging method, and the
kriging with external drift and concluded that the kriging with external drift considering elevation information has the highest accuracy. Martinez-Cob (1996) applied
the co-kriging method to interpolate precipitation in mountainous areas, achieving
great interpolation results. Through the interpolation of a great variety of meteorological elements using the IDW method without trends, the universal kriging method,
and an artificial neural network, Attorre et al. (2007) found that interpolation results
of the universal kriging method are better than those of the other methods. Moreover, many studies have shown that the kriging method outperforms the traditional
methods (Dubrule 1983; Tabios and Salas 1985; Stein and Corsten 1991; Phillips
et al. 1992; Laslett 1994; Burrough and McDonnell 1998; Tan and Ding 2004; Li et al.
2006; Ma et al. 2008; Chu et al. 2008; Zhong 2010; Zhao and Yang 2012). Joly et al.
(2011) compared the differences between the local regression method and the kriging
method in simulating temperature in France, and the results showed that the local
regression method is more suitable for areas with large spatial heterogeneity, while
the kriging method outperforms the local regression method for areas with relatively
small spatial variability since the distance between sampling points greatly affects the
results. Kurtzman and Kadmon (1999) found that for the average climate elements,
the regression method is better than the IDW method and the spline method. In
regions to the east of the Yellow River in Gansu Province, Cai et al. (2009) proposed
a multiple linear regression interpolation method based on DEM that is superior to
the IDW method, the kriging method, and the spline method. In a study on the interpolation of meteorological elements in Canada, Nalder and Wein (1998) showed that
the GIDS method outperforms classical interpolation methods including the kriging
method. Cai et al. (2006) interpolated the precipitation in three provinces in Northeast
China, i.e., Liaoning Province, Jilin Province, and Heilongjiang Province, using the
IDW method, hybrid interpolation (trend surface simulation combined with residual
interpolation), Aunsplin, and the spatial climatic value + interannual anomaly interpolation method, and the results showed that Aunsplin has the best accuracy. By
conducting a spatial interpolation on temperature elements of short time series, Guan
et al. (2007) found that Aunsplin is inferior to the spatial climatic value + interannual anomaly interpolation method. In the terms of the monthly average climatic
values of Canada, Price et al. (2000) compared the interpolation results of Aunsplin
and GIDS, and the results showed that Aunsplin achieves a good simulation on the
climatic values in most months, while the simple and easy-to-understand GIDS is
not suitable for the simulation of precipitation in areas with complex terrain, and the
two methods are likely to produce negative values in simulating climatic values in
areas with sparse stations. By calculating the multi-year average temperature, Lin
et al. (2002) found that GIDS is of great application value in temperature interpolation. Cai et al. (2005) interpolated monthly and annual average temperatures using
the IDW method, the ordinary kriging method, and the trend surface method and

pointed out that the IDW method and the kriging method are applicable to areas with dense stations and flat terrain and that the trend surface method is suitable for spatial interpolation of multi-year average temperature. Some studies have shown that when the meteorological data are densely distributed, differences among the results of different interpolation methods are small (Michaud and Sorooshian 1994; Dirks et al. 1998; Burrough and McDonnell 1998; Hartkamp et al. 1999).

In summary, the review of current studies in China and abroad reveals that the research on spatial interpolation methods has been basically stalled and that it is difficult to develop new methods. The research focus has been shifted from the interpolation method itself to improvement in traditional interpolation methods, and more emphasis is placed on practicability. Some studies have introduced many factors influencing meteorological elements into the analysis; some studies have combined modern mathematical methods with modern scientific and technological methods or traditional methods; and some studies have compared all kinds of traditional methods in specific study areas to obtain the relatively optimal methods.

1.3 Research Objectives, Contents, and Methods

1.3.1 Research Objectives

The objective of this book is to overcome the shortcomings of HASM. According to the principal theorem of surface theory, a more accurate HASM method with a complete theoretical foundation, which is referred to as HASM.MOD, is developed. Then, the quantitative indexes of the stopping criteria for iteration are presented. Considering the disadvantage of HASM being established based on other surface interpolation methods, the sensitivity of HASM and HASM.MOD to the selection of driving field is analysed, and a more robust HASM method independent of the driving field is proposed. Through numerical simulation and case studies, the performance of HASM.MOD with any driving field, especially in a zero driving field, is verified. HASM is completely independent of other interpolation methods and has better practicability. Moreover, the influence of the sampling percentage and sampling error on the accuracy of HASM.MOD is explored to provide references for practical application. Last, to increase the computational speed, parallel computing of HASM.MOD is achieved.

1.3.2   Research Contents and Methods

In this book, by solving the problems currently faced by traditional HASM, the robustness and applicability of HASM are improved, thus providing a more accurate method for spatial interpolation of meteorological elements and other attribute elements. The main research contents are as follows:

1. Based on the principal theorem of surface theory, the traditional HASM is modified, and a stable HASM with a solid theoretical foundation is proposed, referred to as HASM.MOD.

The improved HASM is proposed based on the principle of differential geometry. Traditional HASM mainly applies the differential equations satisfied by the partial derivatives in the x- and y-directions of the Gaussian equations. The mixed partial derivative term is not used since it leads to overdetermined linear algebraic equations or numerical overflow when the central finite difference scheme is used for discretization. Studies have shown that the problem of numerical overflow can be effectively solved by discretizing the differential equations of mixed partial derivatives with the eccentric finite difference scheme. The introduction of differential equations of mixed partial derivatives enables HASM to reflect the principal theorem of surface theory and to more completely utilize the spatial information of the surface, yielding HASM with better numerical stability. In addition, the simulation accuracy of HASM at the boundaries of the computational domains is greatly improved. The influence of the high-order finite difference scheme on the simulation accuracy of HASM is also studied. By comparing the influences of the mixed partial derivative term and the high-order finite difference scheme on the simulation results of HASM, the primary cause of the increase in the accuracy of HASM is discussed. In addition, although the high-order discretization method can improve the accuracy of HASM to some extent, it increases the computational burden of the model.

HASM is divided into two parts, i.e., the inner iteration and outer iteration. In traditional HASM, the number of outer iterations is artificially set based on the experience of the developers, which is subjective, uncertain and without theoretical basis. Based on the principal theorem of surface theory, the quantitative indexes of the stopping criteria for the outer iteration of HASM.MOD are presented, which improves the theoretical basis of HASM and facilitates the intelligent development of HASM.

2. The sensitivity and dependence of HASM on the selection of the driving field are eliminated so that HASM.MOD no longer depends on other interpolation methods

Traditional HASM relies on the other interpolation methods to compute the driving field. If the driving field is randomly selected, the simulation accuracy of HASM is low. Traditional HASM depends on the driving field to varying degrees. Traditional HASM is not completely based on the principle of differential geometry. By introducing the third partial differential equation of the surface into HASM and adopting a stable finite difference discretization scheme, the HASM.MOD has a complete theoretical basis. The sensitivity of HASM.MOD to the selection of driving field is

investigated through numerical simulation, case verification, and theoretical analysis. The improved HASM (HASM.MOD) is expected to achieve a high simulation
accuracy with a randomly selected driving field, especially zero driving field. The
dependence of HASM on the driving field is reduced, the implementation process of
HASM is simplified, and HASM.MOD is completely independent of other methods.

3. The influence of the sampling information on the simulation accuracy of
HASM.MOD is investigated.

HASM takes the partial differential equations as the main equations and the
sampling information as the constraints. Through numerical simulation and case
studies, the influence of the sampling information, especially the sampling percentage
and sampling error, on the simulation results of HASM.MOD is investigated,
providing some references for practical application.

4. The fast computational algorithm of HASM.MOD is studied to increase the
computational speed and to realize parallel computing.

HASM is transformed into the process of solving sparse linear algebraic equations.
Previous experience has shown that the time consumed by solving the linear algebraic
equations accounts for a large proportion of the total computational time. Hence, a
high-speed solving process is particularly important for HASM. By analysing the
structural characteristics of the coefficient matrix of the HASM.MOD equations, the
optimal preconditioning operator based on the CG method is selected to transform the
original equations into well-conditioned equivalent linear equations, thereby accel-
erating the convergence speed and shortening the processing time. On this basis,
with MPI parallel programming, the solving process of HASM is decomposed into
independent multitasks that can be processed in parallel. Multiple processes run in
parallel, with each process executing an independent task, so the equations are solved
quickly and effectively.

5. Some application of surface modelling of climate variables, including precipitation, temperature, potential evapotranspiration, percentage of sunshine and the
application of HASM in downscaling precipitation and temperature in the future.

References

Aalto J, Pirinen P, Heikkinen J, et al. 2012. Spatial interpolation of monthly
Anselin L, Rey S. 1991. Properties of tests for spatial dependence in linear regression models.
    Geographical Analysis, 23(2): 112–131.
Apaydin H, Anli AS, Qzturk F. 2011. Evaluation of topographical and geographical effects on
    some climatic parameters in the Central Anatolia Region of Turkey. International Journal of
    Climatology, 31: 1264–1279.
Ashraf M, Loftis JC, Hubbard KG. 1997. Application of geostatisticals to evaluate partial weather
    station network. Agricultural Forest Meteorology, 84: 255–271.
Attorre F, Alfo M, Sanctis MD, Francesconi F, et al. 2007. Comparison of interpolation methods for
    mapping climatic and bioclimatic variables at regional scale. International Journal of Climatology,
    27(13): 1825–1843.
Bai SR, Li T, Ning JY. 2011. Parallel Kriging on Interpolation of Spatial Precipitation Based on
    MPI. Computing Technology and Automation, 30(1): 71–74.

Band C. 1991. Forest ecosystem processes at the watershed scale: Basis for distributed simulation. Ecological Modeling, 56: 171–196.

Bărbulescu AS, Cristina, Indrecan ML. 2021. Computing the Beta Parameter in IDW Interpolation by Using a Genetic Algorithm. Water, 13(6).

Basist A, Bell GD, Meentemeyer V. 1994. Statistical relationships between topography and precipitation patterns. Journal of climate, 7: 1305–1315.

Bastin G, Lorent B, Duque C, et al. 1984. Optimal estimation of the average rainfall and optimal selection of raingauge locations. Water Resources Research, 20: 463–470.

Bedient PB, Huber WC. 1992. Hydrology and floodplain analysis. New York: Addison.

Benzi R, Deidda R, Marrocu M. 1997. Characterization of temperature and precipitation fields over Sardinia with principal component analysis and singularspectrum analysis. International Journal of Climatology, 17: 1231–1262.

Bloomfield P, Nychka DW. 1992. Climate spectra and detecting climate change. Climate Change, 21: 275–287.

Bloomfield P. 1992. Trends in global temperature. Climate Change, 21: 1–16.

Brunsdon C, Fotheringham AS, Charlton ME. 1996. Geographically weighted regression: A method for exploring spatial nonstationarity. Geographical Analysis, 28(4): 281–298.

Brunsdon C, Mcclatchey J, Unwin DJ. 2001. Spatial variations in the average rainfall-altitude relationship in Great Britian: An approach using geographical weighted regression. International Journal of Climatology, 21: 455–466.

Brus D, Heuvelink G. 2007. Optimization of sample patterns for universial kriging of environmental variables. Geoderma, 138: 86–95.

Burrough PA, McDonnell RA. 1998. Principles of geographical information systems. New York: Oxford University Press.

Burrough PA. 1996. Principles of geographical information systems for land resources assessment. Oxford: Oxford University Press.

Cai DH, Guo N, Li CW. 2009. Research of Spatial Interpolation Methods of Temperature Based on DEM Data. Journal of Arid Meteorology,27(1): 10–17.

Cai F, Yu GR, Zhu QL. 2005. Comparison of Precisions between Spatial Methods of Climatic Factors :A Case Study on Mean Air Temperature. Resources Science, 27(5): 174–179.

Cai F, Yu HB, Jiao LL, et al. 2006.Comparison of Precision of Spatial Interpolation of Precipitation Factors :A Case Study in Northeastern China.Resources Science, 28(6): 73–79.

Caruso C, Quarta F. 1998. Interpolation methods comparison. Computers and Mathematics with Applications, 35(12): 109–126.

Chen CF, Yue TX. 2010a. A method of DEM construction and related error analysis. Computers & Geoscience, 36:717–715.

Chen CF, Yue TX. 2010b. Solving High Accuracy Surface Modelling Based on Preconditioning Conjugate Gradient.Journal o f China University of Mining & Technology, 39(2): 290–294.

Chen CF. 2010. Research on adaptive algorithm of high precision surface modeling. Graduate School of Chinese Academy of Sciences (Institute of Geographical Sciences and resources).

Chen XF, Liu JY, Zhang ZX, et al.1998. Using GIS to Establish Temperature Distribution Model in Mountain Area. Journal of Image and Graphics, 3(3): 234–238.

Chessa PA, DelitalaAM. 1997. Objective analysis of daily extreme temperatures of Sardinia (Italy) using distance from sea as independent variable. International Journal of Climatology, 17: 1467–1485.

Chiles JP. 1999. Geostatistics-modeling spatial uncertainty. A Wiley Interscience Publication.

Chu SL, Zhou ZY, Yuan L, et al. 2008. Study on spatial precipitation interpolation methods—A case of Gansu province. Pratacultura Science, 25(6): 19–23.

climate data for Finland: comparing the performance of kriging and generalized additive models. Theory and Applied Climatology, 112:99–111.

Collins FC, Bolstad PV. 1996. A comparison of spatial interpolation techniques in temperature esti-mation. Proceedings of the Third International Conference on Integrating GIS and Environmental

Modeling, Santa Barbara, CA: National Center for Geographic Information Analysis (NCGIA). CD-ROM.

Courault D, Monestiez P. 1999. Spatial interpolation of air temperature according to atmospheric circulation patterns in southeast France. International Journal of Climatology, 19: 365–378.

Creutin JD, Obled C. 1982. Objective analyses and mapping techniques for rainfall fields: an objective comparison. Water Resources Research, 18 (2): 413–431.

Dale Z, Claire P, Amy R, et al. 1999. An experimental comparison of ordinary and universal kriging and inverse distance weighting, Mathematical Geology, 31(4): 375–390.

Daly C, Kittel G F, McNab A, et al. 2000. Development of a 103-year high-resolution climate data set for the conter minus United S tates1 In Proc1, 12th Conference on Applied Climatology, Asheville, N1C1 8–11 May Boston, Mass1: Am1 Meteorological Society, 249–252.

Daly C, Neilson RP, Phillips DL. 1994. A statistical topographic model for mapping climatological precipitation over montainous terrain. Journal of Applied Meteorology, 33 (2): 140–158.

Daly C, Taylor GH, Gibson WP, et al. 2000b. High-quality spatial climate data sets for the United States and beyond Tansactions of the ASAE, 43(6): 1957–1962.

Deepti H, Chaitra D, Ramesh T, et, al. 2020. Adaptive Cubic Spline Interpolation in CIELAB Color Space for Underwater Image Enhancement. Procedia Computer Science, 171.

Deraisme J, Humbert J, Drogue G, et al. 2001. Geostatistical interpolation of rainfall in mountainous areas, in: Monestiez, P., Allard, D., Froidevaux, R. (Eds.), GeoENV III: Geostatistics for Environmental Applications. Kluwer Academic Publishers, Dordrecht, pp. 57–66.

Dietz T, Rosa EA, York R. 2007. Driving the human ecological footprint. Frontiers in Ecology and the Environment, 5(1), 13–18.

Dingman SL. 1981. Elevation: a major influence on the hydrology of New Hampshire and Vermont, USA. Hydrology Science Bulletin, 26: 399–413.

Dingman SL, Seely-Reynolds DM, Reynolds RC III. 1988. Application of kriging to estimate mean annual precipitation in a region of orographic influence. Water Resources Bulletin, 24: 29–339.

Dingman SL. 2002. Physical hydrology. Prentice Hall, NJ.

Diodato N, Ceccarelli M. 2005. Interpolation process using multivariate geostatistics for mapping of climatological precipitation mean in the Sannil Mountains (southern Italy). Earth Surface processes and Landforms, 30(2): 259–268.

Dirks KN, Hay JE, Stow CD, et al. 1998. High-resolution studies of rainfall on Norfolk Island. Part II: Interpolation of rainfall data. Hydrology, 208: 187–193.

Dobson AP, Bradshaw AD, Baker AJM. 1997. Hopes for the future: Restoration ecology and conservation biology. Science, 277: 515–521.

Dou ZH. 2001. Parallel programming technology of high performance computing MPI parallel programming. Tsinghua university press.

Dubrule O. 1983. Two methods with different objectives: Splines and Kriging. Mathematical Geology, 15: 245–257.

Eckstein BA. 1989. Evaluation of spline and weighted average interpolation algorithm. Computers & Geosciences, 15(1): 79–94.

Eleanor R, Cross MS, Rick Perrine BS. 1984. Predicting areas endemic for schistosomiasis using weather variables and a landsat data base. Mili. Medi. 149(10): 542–545.

Evans IS. 1980. An integrated system of terrain analysis and slope mapping. Zeitschriftfuer Geomorphologie, (36): 274–295.

Fan Y, Bras RI. 1995. On the concept of a representative elementary area in catchments runoff. Hydrology Process, 9(5): 821–932.

Fang JY. 2002. Global Ecology: climate change and ecological response. Higher Education Press.

Felicisimo Perez AM. 1992. El clima de Asturias. Geografia de Asturias 1, 17–32. Prense Iberica.

Feng W, Gao L, Zhou YH, et, al. 2019. Element-free Galerkin scaled boundary method based on moving Kriging interpolation for steady heat conduction analysis. Engineering Analysis with Boundary Elements, 106.

Fernandez Alvarez EM. 1996. Granatlas del Principado de Asturias. Ediciones Nobel, Oviedo.

Fotheringham AS. 2000. Quantitative geography. Sage Publications.

French MN, Krajewski WF, Cuykendal RR. 1992. Rainfall forecasting in space and time using a neural network. Journal of Hydrology, 137: 1–37.

Fu BP. 1983. Mountain climate. Science Press.

Garen DC, Johnson GL, Hanson CL. 1994. Mean areal precipitation for daily hydrologic modeling in mountainous terrain. Water Resources Bulletin, 30: 481–491.

Goodale CL, Alber JD, Ollinger SV. 1998. Mapping monthly precipitation, temperature and solar radiation for Ireland with polynomial regression and digitial elevation model. Climate Research, 10: 35–49.

Goovaerts P. 1997. Geostatistics for Natural Resources Evaluation. New York: Oxford University Press.

Goovaerts P. 1999a. Geostatistics in soil science: state-of-the-art and perspectives. Geoderma, 89, 1–46.

Goovaerts P. 1999b. Performance comparison of geostatistical algorithms for incorporating elevation into the mapping of precipitation. The IV International Conference on Geo Commutation was hosted by Mary Washington College in Fredericksburg, VA, USA, on 25–28 July 1999.

Goovaerts P. 2000. Geostatistical approaches for incorporating elevation into the spatial interpolation of rainfall. Journal of Hydrology, 228: 113–129.

Govindaraju RS, Rao AR. 2000. Neural netaorks in hydrology. Netherlands: Kluwer Academic Publishers.

Grayson R, Bloschl G. 2001. Spatial patterns in catchment hydrology:observaltions and modeling. Cambridge University Press.

Griffiths GA, McSaveney MJ. 1983. Distribution of mean annual precipitation across some steepland regions of New Zealand. New Zealand Journal of Science, 26: 197–209.

Gu KY, Zhou Y, Sun H, et, al. 2020. Spatial distribution and determinants of PM2.5 in China's cities: fresh evidence from IDW and GWR. Environmental monitoring and assessment, 193(1).

Gu ZH, Shi PJ, Chen J. 2006. Precipitation interpolation research over regions with sparse meteorological stations: a case study in Xilin Gole league. Journal of Beijing Normal University (Natural Science), 42(2): 204–208.

Guan HQ, Cai F, Wang Y, et al. 2007. Comparison of different spatial interpolation methods for air temperature data of short-time series. Journal of Meteorology and Environment, 5: 13–16.

Hancock PA, Hutchinson MF. 2006. Saptial interpolation of large climate data sets using bivariate thin plate smoothing splines. Environmental Modeling & Software, 21: 1684–1694.

Hanqin T, Melillo JM, Kicklibhter DW, et al. 1998. Effect of interannual climate variability on carbon storage in Amazonian ecosystems. Nature, 396(6712):664– 667.

Hao J, Liu WJ, Zhang JY, et, al. 2020. Spatiotemporal variations of nitrate sources and dynamics in a typical agricultural riverine system under monsoon climate. Journal of Environmental Sciences, 93.

Hardy RL. 1971. Multiquadric equations of topography and other irregular surfaces. Journal of Geophysical Research, 76(8): 1905–1915.

Hartkamp AD, De Beurs K, Stein A, et al. 1999. Interpolation Techniques for climate variables. NRG-GIS Series 99-01 (Mexico, D.F.: CIMMYT).

Hasenauer H, Merganicova K, Petritsch R, et al. 2003. Validation daily climate interpolations over complex terrain in Austria. Agricultural Forest Meteorology, 119(1–2): 87–107.

Hay L, Viger R, McCabe G. 1998. Precipitation interpolation in mountainous regions using multiple linear regression, in: Kovar, K., Tappeiner, U., Peters, N.E., Craig, R.G. (Eds.), Hydrology, Water Resources and Ecology in Headwaters. Proceedings of the Headwater 1998 Conference, Mean/Merano, Italy, April IAHS Publication 248.

He HY, Guo ZH, Xiao FW. 2005. Research progress of spatial interpolation technology of precipitation. Chinese Journal of Ecology, 24(10): 1187–1191.

Hevesi JA, Istok JD, Flint AL. 1992. Precipitation estimation in mountainous terrain using multivariate geostatistics. Structural analysis. Journal of Applied Meteorology, 31: 661–676.

Hijmans RJ, Cameron S, Parra JL, et al. 2006. Very high resolution interpolated climate surfaces for global land areas. International Journal of Climatology, 25: 1965–1978.

Hjort J, Luoto M. 2010. Novel theoretical insights into geomorphic process-environment relationships using time-series of MODIS LST images. Theoretical and Applied Climatology, 107:
265–277.
Hofierka J, Parajka J, Mitasova M, et al. 2002. Multivariate interpolation of precipitation using
regularized spline with tension. Transactions in GIS, 6: 135–150.
Hofstra N, Haylock M, New M, et al. 2008. Comparison of six methods for the interpolation of
daily, European climate data. Journal of Geographical Research, 113–110.
Holdaway M. 1996. Spatial modeling and interpolation of monthly temperature using kriging.
Climate Research, 6: 215–225.
Hou JR, Huang JX. 1990. Theory and method of Geostatistics. Geological Publishing House.
Hou L, Niu BB, Li XJ, et,al. 2020. Analysis of Soil Fertility Quality and Heavy Metal Pollution in
the Dawen River Basin, China. Open Journal of Soil Science, 10(12).
Houghton JG. 1979. A model for orographic precipitation. Monthly Weather Review, 107: 1463–
1475.
Hulme M, Conway D, Jones PD, et al. 1995. Construction of a 1961–1990 European climatology
for climate change modeling and impace applications. International Journal of Climatology,
15:1333–1363.
Hutchinson MF, Gessler PE. 1994. Splines-more than just a smooth interpolator. Geoderma, 62:
45–67.
Hutchinson MF. 1991. The application of thin plate smoothing splines to continent-wide data
assimilation. In: Jasper JD(ed.) BMRC Research Report No. 27, Data Assimilation Systems.
Melbourne: Bureau of Meteorology, 104–113.
Hutchinson MF. 1995. Interpolating mean rainfall using thin-plate smoothing splines. International
Journal of Geographical Information Systems, 9: 385–403.
Hutchinson MF. 1998a. Interpolation of rainfall data with thin plate smoothing splines-Part I: two
dimensional smoothing of data with short range correlation. Journal of Geographic Information
and Decision Analysis, 2(2): 153–167.
Hutchinson MF. 1998b. Interpolation of rainfall data with thin plate smoothing splines-Part II:
analysis of topographic dependence. Journal of Geographic Information and Decision Analysis
2, 152–167.
Hutchinson MF. 2001. AUNSPLIN user guide. Australian National University,Canberra.
Jarvis C, Stuart N. 2001a. A comparison among strategies for interpolating maximum and minimum
daily air temperatures. Part I: the selection of 'guiding' topographic and land cover. Journal of
Applied Meteorology, 40(6): 1060–1074.
Jarvis C, Stuart N. 2001b. A comparison among strategies for interpolating maximum and minimum
daily air temperatures. Part II: the interaction between number of guiding variables and the type
of interpolation method. Journal of Applied Meteorology, 40(6): 1075–1084.
Jeffrey SJ, Carter JO, Moodie KB, et al. 2001. Using spatial interpolation to construct a
comprehensive archive of Australian climate data. Environmental Modeling & Software, 16:
309–330.
Jiang SH, Ren LL, Yong B, et al.2010.Comparison of spatial interpolation methods for the
precipitation in Laoha River basin. Journal of Arid Land Resources and Environment,24(1):
80–84.
Joly D, Brossard T, Cardot H, et al. 2011. Temperature interpolation based on local information:
the example of France. International Journal of Climatology, 31: 2141–2153.
Journel AG, Huijbregts CJ. 1978. Mining geostatistics. New York: Academic Press.
Kawser A, Kouamana B, Milcade FC. 2019. Spline parameterization based nonlinear trajectory
optimization along 4D waypoints. Advances in Aircraft and Spacecraft Science, 6(5).
Koike K, Matsuda S, Gu B. 2001. Evaluation of interpolation accuracy of Neural Kriging with
application to temperature distribution analysis. Mathematical Geology, 33(4): 421–448.
Krajewski WF. 1987. Co-kriging of radar and rain gauge data. Journal of Geophysics Research,
92(8): 9571–9580.

Kurtzman D, Kadmon R. 1999. Mapping of temperature variables in Israel: a comparison of different interpolation methods. Climate Research, 13: 33–43.

Lam NS. 1983. Spatial interpolation method: a review. American Cartography, 10: 129–149.

Laslett GM. 1994. Kriging and splines: An empirical comparison of their predictive performance in some applications. Journal of the American Statistical Association, 89: 391–400.

Leathwick J, Elith J, Hastie T. 2006. Comparative performance of generalized additive models and multivariate adaptive regression splines for statistical modeling of species distributions. Ecological Modeling, 199: 188–196.

Lennon JJ, Turner JRG. 1995. Predicting the spatial distribution of climate: temperature in Great Britain. Journal of Animal Ecology, 64: 370–392.

Li CK, Chen L, Wang Y. 2007. Research on spatial interpolation of rainfall distribution—— A case study of Idaho State in the USA. Mineral Resources and Geology, 21(6): 684–687.

Li JL, Zhang J, Zhang C, et al. 2006. Analyze and compare the Spatial Interpolation Methods for climate factor. Pratacultural Science,23(8): 6–11.

Li QY, Wang NC, Yi DY. 2006. Numerical analysis. Huazhong University of science and Technology Press

Liao SB, Li ZH. 2004. Some Practical Problems Related to Rasterization of Air Temperature. Meteorological Science and Technology, 32(5) : 352–356.

Lin GF, Chen LH. 2004. A spatial interpolation method based on radial basis function networks incorporating a semivatiogram model. Journal of Hydrology, 288(3–4): 288–298.

Lin ZG. 1995. Topographical Precipitation Climatology.Science Press.

Lin ZH, Mo XG, Li HX, et al. 2002. Comparison of Three Spatial Interpolation Methods for Climate Variables in China. Acta Geographica Sinica,57(1): 47–56.

Lioyd CD. 2005. Assessing the effect of integrating elevation data into the estimation of monthly precipitation in Great Britain. Journal of Hydrology, 308: 128–150.

Liu XA, Yu GR, Fan LS, et al. 2004. Study on spatialization technology of terrestrial eco-information in China:Temperature and precipitation. Journal of Natural Resources, 19(6): 818–825.

Liu CM, Sun R. 1999. Ecological Aspects of Water Cycle: Advances in Soil-Vegetation-Atmosphere of Energy and Water Fuxes. Advances in Water Science,10(3): 251–259.

Liu ZH, Li LT, McVicar TR, et al. 2008. Introduction of the Professional Interpolation Software for Meteorology Data :ANUSPLINN. Meteorological Monthly, 34(2): 92–100.

Liu ZJ, Yu XX, Wang SS, et al. Comparative Analysis of Three Covariates Methods in Thin-Plate Smoothing Splines for Interpolating Precipitation. Progress in Geography, 31(1): 56–62.

Lu GY, Wong DW. 2008. An adaptive inverse-distance weighting spatial interpolation technique. Computers & Geosciences, 34(9): 1044–1055.

Lu YM, Yue TX, Chen CF, et al. 2010. Surface Modelling of Annual Precipitation in China. Journal of Natural Resources, 7:1194–1205.

Ma XL, Li CE, Chen QG. 2008. Study on the method of GIS based spatial interpolation of climate factors in China. Pratacultural Science, 25(11): 13–19.

Maceachren AM, Davidson JV. 1987. Sampling and isometric mapping of continuous geographic surface. The American Cartographer, 14(4): 299–320.

Marquinez J, Lastra J, Garca P. 2003. Estimation models for precipitation inmountainous regions: The use of GIS and multivariate analysis. Journal of Hydrology, 270: 1–11.

Martinez-Cob A. 1996. Multivariate geostatistical analysis of evapotranspiration and precipitation in mountainous terrain. Journal of Hydrology, 174: 19–35.

Masoudi M. 2021. Estimation of the spatial climate comfort distribution using tourism climate index (TCI) and inverse distance weighting (IDW) (case study: Fars Province, Iran). Arabian Journal of Geosciences, 14(5).

Matheron G. 1963. Principles of geostatistics. Economic Geology, 58: 1246–1266.

Matheron G. 1971. The theory of regionalized variables and its applications. Cahiers du Centre de Morphologie Mathematique, Ecole des Mines, Fountainebleau.

McCuen RH. 1998. Hydrologic analysis and design, 2nd ed., Prentice Hall, Englewood Cliff, NJ.

Mebrhatu MT, Tsubo M, Walker S. 2004. A statistical model for seasonal rainfall forecasting over the highlands of Eritrea, in International Crop Science Organization.

Michaud JD, Sorooshian S. 1994. Effect of rainfall-sampling errors on simulations of desert flash floods. Water Resources Reaearch, 39(10): 2765–2775.

Mo L, Zhang QW. 2007.Application of artificial neural network in spatial interpolation of precipitation. Computer and Engineering Institute, 35(9): 9–12.

Nalder IA, Wein RW. 1998. Spatial interpolation of climate normals: Test of a new method in the Canadian Boreal forest. Agricultural and Forest Meteorology, 92: 211–225.

New M, Lister D, Hulme M, et al. 2002. A high resolution data set of surface climate over global land areas. Climate Research, 21: 1–15.

Ninyerola M, Pons X, Roure JM. 2000. A methodological approach of climatological modeling of air temperature and precipitation through GIS techniques. International Journal of Climatology, 20(14): 1823–1841.

Ninyerola M, Pons X, Roure JM. 2007. Objective air temperature mapping for the Iberian Peninsula using spatial interpolation and GIS. International Journal of Climatology, 27:1231–1242.

Oettli P, Camberlin P. 2005. Influence of topography on monthly rainfull distribution over East Africa. Climate Research, 28: 199–212.

Okubo T. 1987. Differential geometry. New York.

Ollinger SV, Aber JD, Lovett GM. 1993. A spatial model of atmospheric deposition for the northeastern U.S. Journal of Applied Ecology, 3: 459–472.

Ollinger SV, Aber JD, Federer CA, et al. 1995. Modeling physical and chemical climatic variables across the Northeastern U.S. for a geographic information system. U.S. Forest Service General Technical Report NE-191, Radnor, PA.

Pan YZ, Gong DY, Deng L, et al. 2004. Smart Distance Searching-based and DEM-informed Interpolation of Surface Air Temprrature in China. Acta Geographica Sinica, 59(3): 366–374.

Perry M. 2006. A spatial analysis of trends in the UK climate since 1914 using gridded datasets. National Climate Information Centre, Climate Memorandum No. 21. pp 29.

Phillips DL, Dolph J, Marks D. 1992. A comparison of geostatistical procedures for spatial analysis of precipitation in mountainous terrain. Agricultural Forest Meteorology, 58: 119–141.

Plantico M, Goss LA, Daly C, et al. 2000. A new U1 S1 climate at las1 in Proc1, 12th Confl on Applied Climatology, Asheville, N1C1 8–11 May 1 Boston, M ass1: Am1 Meteorological Society, 247–248.

Portales C, Boronat N, Pardo-Pascual JE, et al. 2010. Seasonal precipitation interpolation at the Valencia region with multivariate methods using geographic and topographic information. International Journal of Climatology, 30:1547–1563.

Price DT, McKenney DW, Nalder IA, et al. 2000. A comparison of two statistical methods for spatial interpolation of Canadian Monthly mean climate data. Agricultural and Forest Meteorology, 101: 81–94.

Prudhomme C, Duncan WR. 1999. Mapping extreme rainfall in a mountainous region using geostastistical techniques: a case study in Scotland. International Journal of Climatology, 19(12):1337–1356.

Qian YL, Lv HQ, Zhang YH, et al. 2010. Application and assessment of spatial interpolation method on daily meteorological elements based on ANUSPLIN software. Journal of Meteorology and Environment, 26(2): 7–15.

Raveesh G, Goyal R, Tyagi SK. 2021. Advances in atmospheric water generation technologies. Energy Conversion and Management, 239.

Rhind D. 1975. A skeletal overview of spatial interpolation techniques. Computer Application, 2(3–4): 293–309.

Richard F. 1982. Scattered data interpolation: Tests of some methods. Mathematics of Computation, 38(157): 181–199.

Rizzo D, Dougherty D. 1994. Characterization of aquifer properties using artifical neural networks: Neural kriging. Water Resources Research, 30(2): 483–497.

Roe DR, Brooks BR. 2021. Improving the speed of volumetric density map generation via cubic spline interpolation. Journal of molecular graphics & modelling, 104.

Roni A. 1998. Which type of soil vegetation- atmosphere transfer scheme is needed for general circulation models: a proposal for a higher- order scheme. Journal of Hydrology, 212–213: 136–154.

Sato YS, Yamashita F, Sugiura Y, et al. 2004. FIB-assisted TEM study of an oxide array in the root of a friction stirweided aluminum alloy. Scripta Materialia, 50(3): 365– 369.

Schermerhorn VP. 1967. Relations between topography and annual precipitation in western Oregon and Washington. Water Resources Research, 3: 707–711.

Schwarb M. 2001. The Alpine precipitation climate. Evaluation of a high resolution analysis scheme using comprehensive rain-gauge data. Zuercher Klimaschriften.

Seo DJ, 1996. Nolinear estimation of spatial distribution of rainfall-an indicator cokriging approach. Stochastic Hydrology and Hydraulics, 10: 127–150.

Seo DJ, Smith JA. 1993. Rainfall eatimation using rain gauges and radar: A Bayesian approach. Journal of Stochastic Hydrology and Hydraulics, 5(1): 1–14.

Shi WJ, Liu JY, Du ZP, et al. 2009. Surface modeling of soil PH. Geoderma, 150(1–2): 113–119.

Shi WZ. 2010. Principles of modeling uncertainties in spatial data and spatial analyses. CRC Press.

Simanton JR, Osborn HB. 1980. Reciprocal distance estimate of point rainfall. Journal of Hydraulic Engineering Division, 106(7): 1242–1246.

Singh RK, Multari N, Nau-Hix C, et, al. 2019. Rapid Removal of Poly- and Perfluorinated Compounds from Investigation-Derived Waste (IDW) in a Pilot-Scale Plasma Reactor. Environmental science & technology, 53(19).

Smith B, Prentice IC, Sykes MT. 2001. Representation of vegetation dynamics in the modeling of terrestrial ecosystems: comparing two contrasting approaches within European climate space. Global Ecology and Biogeography, 10: 621–637.

Smith RB. 1979. The influence of mountains on the atmosphere. Advances in Geophysics, 21: 87–230.

Solantie R. 1976. The influences of lakes on meso-scale analysis of temperature in Finland. Ilmatieteen Laitoksen tiedonantoja, 30, p130.

Solmaz F, Ali RV, Seyed KA, et, al. 2020. Comparison of spectral and spatial-based approaches for mapping the local variation of soil moisture in a semi-arid mountainous area. Science of the Total Environment, 724.

Somasundaram D. 2005. Differential geometry. Harrow, U.K.: Alpha Science International Ltd, 382–405.

Song YJ, Yue TX. 2009. High Accuracy Surface Modeling Based on Multi-grid Method. Geomatics and Information Science of Wuhan University, 34(6): 711–714.

Stein A, Corsten LCA. 1991. Universial kriging and cokriging as regression procedures. Biometrics, 47: 575–587.

Su BQ, Hu HS. 1979. Differential geometry. People's Education Press.

Sun RH, Liu QL, Chen LD. 2010. Study on Precipitation Interpolation Based on the Geostatistical Analyst Method. Journal of China Hydrology, 30(1): 14–18.

Szentimrey T, Bihari Z, Szalai S. 2007. Comparison of geostatistical and meteorological interpolation methods. In: Dobesh H et al. (Ed.) Spatialinterpolation for climate data: the use of GIS in climatology and meteorology pp.45–56. London: ISTE Ltd.

Tabios GQ, Salas JD. 1985. A comparative analysis of techniques for spatialinterpolation of precipitation. Water Resources Bulletin, 21(3): 365–380.

Tan JQ, Ding MZ. 2004. An Evaluation of Spatial Data Interpolation Methods. Geomatics and Spatial Information Technology, 27(4): 11–13.

Tang YC. 1985. The distribution of precipitation in mountain QiLian(NANSHAN). Acta Geographica Sinica, 40(4): 323–332.

Teegavarapu RSV, Chandramouli V. 2005. Improved weighting methods,deterministic and stochastic data-driven models for estimation of missing precipitation records. Journal of Hydrology, 312: 191–206.

Teegavarapu RSV. 2007. Use of universal function approximation in variance dependent interpolation technique: an application in Hydrology. Journal of Hydrology, 332: 16–29.

Teegavarapu RSV. 2009. Estimation of missing precipitation records integrating surface interpolation techniques and spatio-temporal association rules. Journal of Hydroinformatics, 11(2):133–146.

Tiessen AH. 1911. Precipitation averages for large areas. Monthly Weather Review, 39: 1082–1084.

Tobin C, Nicotina L, Parlange MB, et al. 2011. Improved interpolation of meteorological forcings for hydrologic applications in a Swiss Alpine region. Journal of Hydrology, 401: 77–89.

Tomczak M. 1998. Spatial interpolation and its uncertainty using automated anisotropic inverse distance weighting (IDW) cross validation/jackknife approach. Journal of Geographic Information and Decision Analysis, 2: 18–30.

Tong ST, Naramngam S. 2007. Modeling the impact of farming practices on water quality in the Little Miami River Basin. Environmental Management, 39: 853–866.

Toponogov VA. 2006. Differential geometry of curves and surfaces. New York: Birkhaeuser Boston.

Uribe A, Vial J, Mauritsen T. 2021. Sensitivity of Tropical Extreme Precipitation to Surface Warming in Aquaplanet Experiments Using a Global Nonhydrostatic Model. Geophysical Research Letters, 48(9).

Vajda A, Venalainen A. 2003. The influence of natural conditions on the spatial variation of climate in Lapland, northern Finland. International Journal of Climatology, 23: 1011–1022.

Venier LA, Hopkin AA, McKenney DW, et al. 1998a. A spatial climate determined risk rating for Scleroderris disease of pines in Ontario. Canadian Jouranl of Forest Research, 28: 1398–1404.

Venier LA, McKenney DW, Wang Y, et al. 1998b. Models of large-sacle breeding-bird distribution as a function of macroclimate in Ontario, Canada. Jouranl of Biogeography, 26: 315–328.

Vieux BE. 2001. Distributed hydrologic modeling using GIS. Water Science and Technology Library. Kluwer Academic Publishers.

Vogt JV, ViauAA, Paquet F. 1997. Mapping regional air temperature fields using satellite-derived surface skin temperatures. International Journal of Climatology, 17: 1559–1579.

Wahba G, Wendelberger J. 1980. Some new mathematical methods for variational objective analysis splines and cross validation. Monthly Weather Review, 108: 1122–1143.

Wang GX, Cheng GG. 1998. The spatial differential features of eco-environment in inland river basins.Scientia Geographica Sinica, 18(4): 355–361.

Wang S, Yan DH, Qin TL, et al. 2011a. Spatial interpolation of precipitation using the PER-Kriging method. Advances in Water Science, 22(6): 756–763.

Wang YX, Gang H, Tao WG, et, al. 2020. Improved RSS Data Generation Method Based on Kriging Interpolation Algorithm. Wireless Personal Communications, 115(prepublish).

Wang Z, Wu YJ, Liang FC, et al. 2011b. Research on spatial interpolation method of annual precipitation in Xinjiang. Agrometeorology in China, 32(3): 331–337.

Wang ZL, Chen XH, Liu DD. 2007. Interpolation of annual average rainfall based on BP neural network method. Rural water conservancy and hydropower in China, 1(3): 57–61.

Wei TC, McGuinness JL. 1973. Reciprocal distance squared method, a computer technique for estimating area precipitation. Technical Report ARS-Nc-8., US Agricultural Research Service, North Central Region, Ohio.

Weisse AK, Bois P. 2001. Topographic effects on statistical characteristics of heavy rainfall and mapping in the French. Journal of Applied Meteorology, 40(4): 720–740.

Weng QH. 2006. An evaluation of spatial interpolation accuracy of elevation data. Progress in Spatial Data Handling, 805:824.

Willmott CJ, Mastsuura K. 1995. Smart interpolation of annually averaged air temperature in the United Stateds. Journal of Applied Meteorology, 34: 2577–2586.

Willmott CJ, Robeson SM. 1995. Climatologically aided interpolation of terrestrial air temperature. International Journal of Climatology, 15(2): 221–229.

Woodward WA, Gray HL. 1993. Global warming and the problem of testing for trend in time series data. Journal of Climate, 6: 953–962.

Wotling G, Bouvier CH, Danloux J. et al. 2000. Regionalization of extreme precipitation distribution using the principal components of the topographical environment. Journal of Hydrology, 233: 86–101.

Xiong B, Li RP, Johnson D, et, al. 2021. Spatial distribution, risk assessment, and source identification of heavy metals in water from the Xiangxi River, Three Gorges Reservoir Region, China. Environmental geochemistry and health, 43(2).

Xu C, Wu DQ, Zhang ZG. 2008. Comparative study of spatial interpolation methods on weather data in Shandong Province. Journal of Shandong University(Natural Science), 43(3):1–5.

Yan H, Nix HA, Hutchinson MF, et al. 2005. Spatial interpolation of monthly mean climate data for China. International Journal of Climatology, 25: 1369–1379.

Yan H. 2004. Thin plate smoothing spline interpolation and climate spatial simulation in China.Scientia Geographica Sinica, 249(2): 163–169.

Yang FM, Sun YK, Yu TY, et al. 2009. Analysis of temporal and spatial variation of air temperature in Heilongjiang Province in recent 10 years. Journal of GEO-Information Science, 11(5): 585–596.

You SC, Li J. 2005. Influence of altitude error on spatial interpolation error of air temperature. Journal of Natural Resources, 20(1):140–144.

Yue TX, Ai NS. 1990. Mathematical model of glacier morphology. Glacial permafrost, 12(3): 227–234.

Yue TX, Chen CF, Li BL. 2012. An adaptive method of high accuracy surface modeling and its application to simulating elevation surface. Transactions inGIS, 14(5): 615–630.

Yue TX, Chen SP, Xu B, et al. 2002. A curve-theorem based approach for change detection and its application to Yellow River delta. International Journal of Remote Sensing, 23(11): 2283–2292.

Yue TX, Du ZP, Liu JY. 2004. High precision surface modeling and error analysis. Progress in Natural Science, 14(3): 300–306.

Yue TX, Du ZP, Song DJ, et al. 2007a. A new method of high accuracy surface modeling and its application to DEM construction. Geomorphology, 91: 161–172.

Yue TX, Du ZP, Song DJ. 2007b. Improvement of high precision surface modeling: HASM4.Journal of image and graphics, 12(2): 343–348.

Yue TX, Du ZP, Song YJ. 2008. Ecological models: spatial models and geographic information system. In: Jorgensen, Sven Erik, Fath, Brian (Eds.), Encyclopedia of Ecology. Elsevier Limited, England, 3315–3325.

Yue TX, Du ZP. 2005. High precision surface modeling: the core module of new generation GIS and CAD. Progress in Natural Science, 15(3): 73–82.

Yue TX, Du ZP. 2006a. Numerical experimental analysis of the best expression of high precision surface modeling. Geo Information Science, 8(3): 83–87.

Yue TX, Du ZP. 2006b. Error comparison between high precision surface modeling and classical model. Progress in Natural Science, 16(8): 986–991.

Yue TX, Fan ZM, Liu JY. 2005. Change of major terrestrial ecosystems in China since 1960. Global and Planetary Change, 48: 287–302.

Yue TX, Fan ZM, Liu JY. 2007. Scenarios of land cover in China. Global and Planetary Change, 55: 317–342.

Yue TX, Li QQ. 2010. Relationship between species diversity and ecotype diversity. Annals of the New York Academy of Sciences 1195: E40–E51.

Yue TX, Song DJ, Du ZP, et al. 2010a. High accuracy surface modeling and its application to DEM generation. International Journal of Remote Sensing, 31(8):2205–2226.

Yue TX, Song YJ. 2008. The Yue–HASM Method. In proceeding of the 8th international symposium on spatial accuracy assessment in natural resources and environmental sciences Y G a G M F Deren Li, ed. (Shanghai), pp. 148–153.

Yue TX, Wang Q, Lu YM, et al. 2010b. Change trends of food provisions in China.Global Planet Change, 72: 118–130.

Yue TX, Wang SH. 2010. Adjustment computation of HASM: a high-accuracy and high-speed method. International Journal of Geographical Information Science 24(11): 1725–1743.

Yue TX, Zhao N, Ramsey RD, et al. 2013a. Climate change trend in China, with improved accuracy. Climatic Change, 120: 137–151.

Yue TX, Zhao N, Yang H, et al. 2013b. A multi-grid method of high accuracy surface modeling and its validation. Transctions in GIS,17(6): 943–952.

Yue TX. 2011. Surface modeling: high accuracy and high speed methods. CRC Press.

Yue WZ, Xu JH, Xu LH. 2005b. Spatial interpolation of climate elements based on Geostatistics. Plateau Meteorology, 24(6): 974–979.

Zeng HE, Huang SH. 2007. Research on spatial data interpolation based on Kriging interpolation. Engineering of Surveying and Mapping,16(5): 5–8.

Zhang LJ, Gove JH, Heath LS. 2005. Spatial residual analysis of six modeling techniques. Ecological Modelling, 186: 154–177.

Zhang S, Liao SB. 2011.Simulation analysis of multi-year mean temperature spatial BP neural network model. Journal of GEO-Information Science, 13(4): 534–538.

Zhang XC, Yang ZN. 1991. Preliminary analysis of water balance in Binggou watershed of Qilian Mountain. Glacial permafrost, 13(1): 35–42.

Zhao CY, Feng ZD, Nan ZR. 2008. Modeling the temporal and spatial variabilities of precipitation in Zulihe River Basin of the Western Loess Plateau. Plateau Meteorology, 27(1): 208–214.

Zhao DZ, Zhang WC, Liu SC. 2004. Research on prism spatial interpolation of geographical features based on DEM. Geographical Science, 24(2): 205–211.

Zhao N, Yue TX. 2014. A modification of HASM for interpolating precipitation in China. Theoretical and Applied Climatology, 116: 273–285.

Zhao T, Yang XY. 2012. Research on Spatial Interpolation Methods of Annual Average Precipitation on Loess Plateau . Ground water, 34(2):189–191.

Zheng X, Basher RE, Thompson CS. 1997. Trend detection in regional-mean temperature series: maximum, minimum, mean, diurnal range, and SST. Journal of Climate, 10: 317–0326.

Zheng X, Basher RE. 1998. Structural time series models and trend location in global and regional temperature series. Journal of Climate, 12: 2347–2358.

Zhong JL. 2010. Study on Spatial Precipitation Interpolation Precision Based on GIS in Xinjiang. Arid Environmental Monitoring, 24(1): 43–57.

Zhu HZ, Luo TX, Daly C. 2003. Validation of simulated grid data sets of China' s temperature and precipitation with high spatial resolution. Geographical Research, 22(3): 349–359.

Zhu QA, Jiang H, Song XD. 2009. Simulation and Analysis of Spatial-temporal Patterns of Acid Rain in Southern China Based on Spatial Interpolation. Research of Environmental Sciences, 22(11): 1237–1244.

Zhu QA, Zhang WC, Zhao DZ. 2005a. Topography-based spatial daily precipitation interpolation by means of PRISM and Thiessen Polygon analysis. Scientia Geographica Sinica, 25(2): 233–238.

Zhu QL, Zhang LZ, Yu GR, et al. 2005b. The spatial and temporal variability characteristics of precipitation in the Yellow River Basin of recent 30 years. Journal of Natural Resources, 20(5): 60–66.

Chapter 2
Modern HASM Method

This chapter focuses improving traditional high-accuracy surface modelling (HASM) and proposes a more accurate HASM method with a complete theoretic basis—the modern HASM method (HASM.MOD). The main reasons for the improved accuracy of HASM are discussed, and the quantitative index of the stopping criteria for iteration in HASM is given.

2.1 Improvements in Traditional HASM Methods

According to the principal theorem of surface theory (Su and Hu 1979; Somasundaram 2005), when the first and second fundamental coefficients of a surface E, F, G, L, M, N satisfy symmetry, E, F, G are positive definite, and E, F, G, L, M, N satisfy the Gauss-Codazzi equations, then the total differential Eq. (2.1) has a unique solution $z = f(x, y)$ under the initial condition of $f(x, y) = f(x_0, y_0)$ $(x = x_0, y = y_0)$.

$$\begin{cases} f_{xx} = \Gamma_{11}^1 f_x + \Gamma_{11}^2 f_y + \frac{L}{\sqrt{E+G-1}} \\ f_{yy} = \Gamma_{22}^1 f_x + \Gamma_{22}^2 f_y + \frac{N}{\sqrt{E+G-1}} \\ f_{xy} = \Gamma_{12}^1 f_x + \Gamma_{12}^2 f_y + \frac{M}{\sqrt{E+G-1}} \end{cases} \tag{2.1}$$

where $E = 1 + f_x^2, F = f_x \cdot f_y, G = 1 + f_y^2$,

$L = \frac{f_{xx}}{\sqrt{1+f_x^2+f_y^2}}, M = \frac{f_{xy}}{\sqrt{1+f_x^2+f_y^2}}, N = \frac{f_{yy}}{\sqrt{1+f_x^2+f_y^2}}$,

$\Gamma_{11}^1 = \frac{1}{2}\left(GE_x - 2FF_x + FE_y\right)\left(EG - F^2\right)^{-1}$,

$\Gamma_{11}^2 = \frac{1}{2}\left(EF_x - EE_y - FE_x\right)\left(EG - F^2\right)^{-1}$,

$\Gamma_{22}^1 = \frac{1}{2}\left(2GF_y - GG_x - FG_y\right)\left(EG - F^2\right)^{-1}$,

$\Gamma_{22}^2 = \frac{1}{2}\left(EG_y - 2FF_y + FG_x\right)\left(EG - F^2\right)^{-1}$,

$$\Gamma_{12}^1 = \tfrac{1}{2}\left(GE_y - FG_x\right)\left(EG - F^2\right)^{-1},$$
$$\Gamma_{12}^2 = \tfrac{1}{2}\left(EG_x - FE_y\right)\left(EG - F^2\right)^{-1}.$$

The traditional HASM model performs a finite-difference discretization of the set of Gauss equations after removing the mixed partial derivatives; that is, numerical simulation of Eq. (2.2) is performed.

$$\begin{cases} f_{xx} = \Gamma_{11}^1 f_x + \Gamma_{11}^2 f_y + \dfrac{L}{\sqrt{E+G-1}} \\ f_{yy} = \Gamma_{22}^1 f_x + \Gamma_{22}^2 f_y + \dfrac{N}{\sqrt{E+G-1}} \end{cases} \qquad (2.2)$$

Although traditional HASM has higher simulation accuracy than the classical interpolation methods, there are individual cases where the interpolation of the HASM model is not ideal in practice. The main reason may be the poor accuracy of the finite-difference discretization described in Eq. (2.2), as well as the fact that the traditional HASM considers only the first two equations in Eq. (2.1) and discards the equation satisfied by the mixed partial derivative terms. Nonetheless, the main reason for error in the HASM model is still unclear. Regarding the oscillating boundary in the simulation region, traditional HASM still does not completely solve this problem, and the simulation values at the boundary are provided by other interpolation methods, with a low accuracy. Although the traditional HASM model is based on the principal theorem of surface theory, its theoretical foundation is still incomplete. Since the principal theorem of surface theory is based on the three Gauss equations, the third equation involves the second type of fundamental coefficient M of the surface, which is related to the local curvature of the surface in space and inevitably has an impact on the shape of the surface. Since the traditional HASM discards the third equation, its theoretical basis is not perfect, which contributes to the non-ideal simulation results.

For Eq. (2.1), the high-order discretization of f_x, f_{xx} is as follows.

$$(f_x)_{(i,j)} \approx \begin{cases} \dfrac{-3f_{0,j}+4f_{1,j}-f_{2,j}}{2h} & i = 0 \\ \dfrac{f_{i+1,j}-f_{i-1,j}}{2h} & i = 1, \dots I, \\ \dfrac{3f_{I+1,j}-4f_{I,j}+f_{I-1,j}}{2h} & i = I+1 \end{cases}$$

$$(f_{xx})_{(i,j)} \approx \begin{cases} \dfrac{2f_{0,j}-5f_{1,j}+4f_{2,j}-f_{3,j}}{h^2} & i = 0, 1 \\ \dfrac{-f_{i+2,j}+16f_{i+1,j}-30f_{i,j}+16f_{i-1,j}-f_{i-2,j}}{12h^2} & i = 2, \dots I-1, \\ \dfrac{2f_{I+1,j}-5f_{I,j}+4f_{I-1,j}-f_{I-2,j}}{h^2} & i = I, I+1 \end{cases} \qquad (2.3)$$

Similarly, the high-order discretization of f_y, f_{yy} is as follows.

$$(f_y)_{(i,j)} \approx \begin{cases} \dfrac{-3f_{i,0}+4f_{i,1}-f_{i,2}}{2h} & j = 0 \\ \dfrac{f_{i,j+1}-f_{i,j-1}}{2h} & j = 1, \dots J, \\ \dfrac{3f_{i,J+1}-4f_{i,J}+f_{i,J-1}}{2h} & j = J+1 \end{cases}$$

$$(f_{yy})_{(i,j)} \approx \begin{cases} \frac{2f_{i,0}-5f_{i,1}+4f_{i,2}-f_{i,3}}{h^2} & j = 0, 1 \\ \frac{-f_{i,j+2}+16f_{i,j+1}-30f_{i,j}+16f_{i,j-1}-f_{i,j-2}}{12h^2} & j = 2, \ldots J - 1, \\ \frac{2f_{i,J+1}-5f_{i,J}+4f_{i,J-1}-f_{i,J-2}}{h^2} & j = J, J + 1 \end{cases} \tag{2.4}$$

During the early stages of HASM, Yue (2011) considered all Gaussian equations; the discretization of the mixed partial derivative terms inside the region is expressed as

$$(f_{xy})_{(i,j)} = \frac{f_{i+1,j+1} - f_{i-1,j+1} + f_{i-1,j-1} - f_{i+1,j-1}}{4h^2}, i = 1, \cdots, I, j = 1, \cdots, J$$

However, when the third equation (cross term equation) is used in the model, data overflow occurs during the calculation, which leads to calculation termination. The overflow is mainly due to the lack of information about the discretization points in the finite-difference discretization of the mixed partial derivative terms, which leads to a lack of diagonal dominance in the coefficient matrix of the ultimately obtained set of equations by discretization and thus introduces numerical overflow during computation (Karniadakis and Kirby 2003).

In this chapter, the mixed partial derivative terms are discretized as follows:

$$(f_{xy})_{(i,j)} \approx \begin{cases} \frac{f_{1,1}-f_{1,0}-f_{0,1}+f_{0,0}}{h^2} & i = 0, j = 0 \\ \frac{f_{1,J+1}+f_{0,J}-f_{1,J}-f_{0,J+1}}{h^2} & i = 0, j = J + 1 \\ \frac{f_{1,j+1}-f_{0,j+1}+f_{0,j-1}-f_{1,j-1}}{2h^2} & i = 0, j = 1, \cdots, J \\ \frac{f_{I+1,1}-f_{I,0}-f_{I,1}+f_{I+1,0}}{h^2} & i = I + 1, j = 0 \\ \frac{f_{I,J}-f_{I+1,J}-f_{I,J+1}+f_{I+1,J+1}}{h^2} & i = I + 1, j = J + 1 \\ \frac{f_{I+1,j+1}-f_{I,j+1}+f_{I,j-1}-f_{I+1,j-1}}{2h^2} & i = I + 1, j = 1, \cdots, J \\ \frac{f_{i+1,1}-f_{i+1,0}+f_{i-1,0}-f_{i-1,1}}{2h^2} & i = 1, \cdots, I, j = 0 \\ \frac{f_{i+1,J+1}-f_{i+1,J}+f_{i-1,J}-f_{i-1,J+1}}{2h^2} & i = 1, \cdots, I, j = J + 1 \\ \frac{f_{i+1,j+1}-f_{i+1,j}-f_{i,j+1}+2f_{i,j}-f_{i-1,j}-f_{i,j-1}+f_{i-1,j-1}}{2h^2} & i = 1, \cdots, I, j = 1, \cdots, J \end{cases}$$

The discretization at the grid point (i,j) fully uses the information at (i,j). Thus, the coefficient matrix of the final set of algebraic equations has a structure that facilitates further improvement in the interpolation accuracy of the HASM model.

Based on the discretization, the equation set to be solved by HASM.MOD is

$$\begin{cases} \frac{-f_{i+2,j}^{n+1}+16f_{i+1,j}^{n+1}-30f_{i,j}^{n+1}+16f_{i-1,j}^{n+1}-f_{i-2,j}^{n+1}}{12h^2} = (\Gamma_{11}^1)_{i,j}^n \frac{f_{i+1,j}^n-f_{i-1,j}^n}{2h} + (\Gamma_{11}^2)_{i,j}^n \frac{f_{i,j+1}^n-f_{i,j-1}^n}{2h} + \frac{L_{ij}^n}{\sqrt{E_{i,j}^n+G_{i,j}^n-1}} \\ \frac{-f_{i,j+2}^{n+1}+16f_{i,j+1}^{n+1}-30f_{i,j}^{n+1}+16f_{i,j-1}^{n+1}-f_{i,j-2}^{n+1}}{12h^2} = (\Gamma_{22}^1)_{i,j}^n \frac{f_{i+1,j}^n-f_{i-1,j}^n}{2h} + (\Gamma_{22}^2)_{i,j}^n \frac{f_{i,j+1}^n-f_{i,j-1}^n}{2h} + \frac{N_{ij}^n}{\sqrt{E_{i,j}^n+G_{i,j}^n-1}} \\ \frac{f_{i+1,j+1}^{n+1}-f_{i+1,j}^{n+1}-f_{i,j+1}^{n+1}+2f_{i,j}^{n+1}-f_{i-1,j}^{n+1}-f_{i,j-1}^{n+1}+f_{i-1,j-1}^{n+1}}{2h^2} = (\Gamma_{12}^1)_{i,j}^{(n)} \frac{f_{i+1,j}^n-f_{i-1,j}^n}{2h} + (\Gamma_{12}^2)_{i,j}^n \frac{f_{i,j+1}^n-f_{i,j-1}^n}{2h} + \frac{M_{ij}}{\sqrt{E_{i,j}^n+G_{i,j}^n-1}} \end{cases}$$

where,

where,

$$E_{i,j}^n = 1 + \left(\frac{f_{i+1,j}^n - f_{i-1,j}^n}{2h}\right)^2, F_{i,j}^n = \left(\frac{f_{i+1,j}^n - f_{i-1,j}^n}{2h}\right)\left(\frac{f_{i,j+1}^n - f_{i,j-1}^n}{2h}\right), G_{i,j}^n = 1 + \left(\frac{f_{i,j+1}^n - f_{i,j-1}^n}{2h}\right)^2,$$

$$L_{i,j}^n = \frac{\frac{f_{i-1,j}^n - 2f_{i,j}^n + f_{i+1,j}^n}{2h^2}}{\sqrt{1 + \left(\frac{f_{i+1,j}^n - f_{i-1,j}^n}{2h}\right)^2 + \left(\frac{f_{i,j+1}^n - f_{i,j-1}^n}{2h}\right)^2}},$$

$$M_{i,j}^n = \frac{\frac{f_{i+1,j+1}^n - f_{i+1,j}^n - f_{i,j+1}^n + 2f_{i,j}^n - f_{i-1,j}^n - f_{i,j-1}^n + f_{i-1,j-1}^n}{2h^2}}{\sqrt{1 + \left(\frac{f_{i+1,j}^n - f_{i-1,j}^n}{2h}\right)^2 + \left(\frac{f_{i,j+1}^n - f_{i,j-1}^n}{2h}\right)^2}},$$

$$N_{i,j}^n = \frac{\frac{f_{i,j-1}^n - 2f_{i,j}^n + f_{i,j+1}^n}{2h^2}}{\sqrt{1 + \left(\frac{f_{i+1,j}^n - f_{i-1,j}^n}{2h}\right)^2 + \left(\frac{f_{i,j+1}^n - f_{i,j-1}^n}{2h}\right)^2}},$$

$$\left(\Gamma_{11}^1\right)_{i,j}^n = \frac{G_{i,j}^n(E_{i+1,j}^n - E_{i-1,j}^n) - 2F_{i,j}^n\left(F_{i+1,j}^n - F_{i-1,j}^n\right) + F_{i,j}^n\left(E_{i,j+1}^n - E_{i,j-1}^n\right)}{4h\left(E_{i,j}^n G_{i,j}^n - (F_{i,j}^n)^2\right)},$$

$$\left(\Gamma_{11}^2\right)_{i,j}^n = \frac{2E_{i,j}^n\left(F_{i+1,j}^n - F_{i-1,j}^n\right) - E_{i,j}^n(E_{i,j+1}^n - E_{i,j-1}^n) - F_{i,j}^n\left(E_{i,j+1}^n - E_{i,j-1}^n\right)}{4h\left(E_{i,j}^n G_{i,j}^n - (F_{i,j}^n)^2\right)},$$

$$\left(\Gamma_{22}^1\right)_{i,j}^n = \frac{2G_{i,j}^n\left(F_{i,j+1}^n - F_{i,j-1}^n\right) - G_{i,j}^n(G_{i+1,j}^n - G_{i-1,j}^n) - F_{i,j}^n\left(G_{i,j+1}^n - G_{i,j-1}^n\right)}{4h\left(E_{i,j}^n G_{i,j}^n - (F_{i,j}^n)^2\right)},$$

$$\left(\Gamma_{22}^2\right)_{i,j}^n = \frac{E_{i,j}^n\left(G_{i,j+1}^n - G_{i,j-1}^n\right) - 2F_{i,j}^n(F_{i,j+1}^n - F_{i,j-1}^n) + F_{i,j}^n\left(G_{i+1,j}^n - G_{i-1,j}^n\right)}{4h\left(E_{i,j}^n G_{i,j}^n - (F_{i,j}^n)^2\right)},$$

$$\left(\Gamma_{12}^2\right)_{i,j}^n = \frac{E_{i,j}^n\left(G_{i+1,j}^n - G_{i-1,j}^n\right) - F_{i,j}^n(E_{i,j+1}^n - E_{i,j-1}^n)hh}{4h\left(E_{i,j}^n G_{i,j}^n - (F_{i,j}^n)^2\right)},$$

$$\left(\Gamma_{12}^1\right)_{i,j}^n = \frac{G_{i,j}^n\left(E_{i+1,j}^n - E_{i-1,j}^n\right) - F_{i,j}^n(G_{i+1,j}^n - G_{i-1,j}^n)hh}{4h\left(E_{i,j}^n G_{i,j}^n - (F_{i,j}^n)^2\right)}.$$

To ensure the simulation accuracy at the sampling points, the HASM model has to satisfy: $f_{i,j} = \tilde{f}_{i,j}, (x_i, y_j) \in \Phi$, where $\Phi = \left\{\left(x_i, y_j, \tilde{f}_{i,j}\right) | 0 \leq i \leq I + 1, 0 \leq j \leq J + 1\right\}$ is the set of sampling points. The matrix expression of the above differential equations is:

$$\begin{cases} \bar{\mathbf{A}}z^{n+1} = \bar{\mathbf{d}}^n \\ \bar{\mathbf{B}}z^{n+1} = \bar{\mathbf{q}}^n \\ \bar{\mathbf{C}}z^{n+1} = \bar{\mathbf{p}}^n \end{cases} \qquad (2.5)$$

where

$$\bar{A} = \begin{bmatrix} 2I_{(J+2)\times(J+2)} & -5I_{(J+2)\times(J+2)} & 4I_{(J+2)\times(J+2)} & -I_{(J+2)\times(J+2)} & & & \\ & 2I_{(J+2)\times(J+2)} & -5I_{(J+2)\times(J+2)} & 4I_{(J+2)\times(J+2)} & -I_{(J+2)\times(J+2)} & & \\ -I_{(J+2)\times(J+2)} & 16I_{(J+2)\times(J+2)} & -30I_{(J+2)\times(J+2)} & 16I_{(J+2)\times(J+2)} & -I_{(J+2)\times(J+2)} & & \\ & \ddots & \ddots & \ddots & \ddots & \ddots & \\ & & -I_{(J+2)\times(J+2)} & 16I_{(J+2)\times(J+2)} & -30I_{(J+2)\times(J+2)} & 16I_{(J+2)\times(J+2)} & -I_{(J+2)\times(J+2)} \\ & & & -I_{(J+2)\times(J+2)} & 4I_{(J+2)\times(J+2)} & -5I_{(J+2)\times(J+2)} & 2I_{(J+2)\times(J+2)} \\ & & & & -I_{(J+2)\times(J+2)} & 4I_{(J+2)\times(J+2)} & -5I_{(J+2)\times(J+2)} & 2I_{(J+2)\times(J+2)} \end{bmatrix}$$

$$\bar{\mathbf{B}} = \begin{bmatrix} \overline{\mathbf{B}_{(J+2)\times(J+2)}} & & \\ & \ddots & \\ & & \overline{\mathbf{B}_{(J+2)\times(J+2)}} \end{bmatrix},$$

$$\overline{\mathbf{B}_{(J+2)\times(J+2)}} = \begin{bmatrix} 2 & -5 & 4 & -1 & & & \\ & 2 & -5 & 4 & -1 & & \\ -1 & 16 & -30 & 16 & 1 & & \\ & \ddots & \ddots & \ddots & \ddots & \ddots & \\ & & -1 & 16 & -30 & 16 & 1 \\ & & & -1 & 4 & -5 & 2 \\ & & & & -1 & 4 & -5 & 2 \end{bmatrix}.$$

$$\bar{\mathbf{C}} = \begin{bmatrix} C_1 & C_2 & & & \\ C_3 & C_4 & C_2 & & \\ & \ddots & \ddots & \ddots & \\ & & C_3 & C_4 & C_2 \\ & & & C_1 & C_2 \end{bmatrix}_{(I+2)\cdot(J+2)\times(I+2)\cdot(J+2)},$$

$$\mathbf{C}_1 = \begin{bmatrix} 1 & -1 & & & \\ & 1 & -1 & & \\ & & \ddots & \ddots & \\ & & & 1 & -1 \\ & & & & 1 & -1 \end{bmatrix}_{(J+2)\times(J+2)},$$

$$\mathbf{C}_2 = \begin{bmatrix} -1 & 1 & & & \\ & -1 & 1 & & \\ & & \ddots & \ddots & \\ & & & -1 & 1 \\ & & & & -1 & 1 \end{bmatrix}_{(J+2)\times(J+2)},$$

$$\mathbf{C}_3 = \begin{bmatrix} 1 & -1 & & & \\ 1 & \text{-}1 & & & \\ & \ddots & \ddots & & \\ & & 1 & \text{-}1 & \\ & & & 1 & -1 \end{bmatrix}_{(J+2)\times(J+2)} ,$$

$$\mathbf{C}_4 = \begin{bmatrix} 0 & 0 & 0 & & \\ -1 & 2 & -1 & & \\ & \ddots & \ddots & \ddots & \\ & & -1 & 2 & -1 \\ & & & 0 & 0 & 0 \end{bmatrix}_{(J+2)\times(J+2)} .$$

Eventually, HASM.MOD solves the following least-squares problem:

$$\begin{cases} min \left\| \begin{bmatrix} \bar{A} \\ \bar{B} \\ \bar{C} \end{bmatrix} z^{n+1} - \begin{bmatrix} \mathbf{d} \\ \mathbf{q} \\ \mathbf{p} \end{bmatrix}^n \right\|_2 \\ \text{s.t. } \mathbf{S}z^{n+1} = k \end{cases} \tag{2.6}$$

where $\mathbf{S}$ and $\mathbf{k}$ are the coefficient matrix of the sampling points and the values of the sampling points, similar to the traditional HASM. Through the introduction of the weight parameter λ, the above least-squares problem is transformed into an optimization problem:

$$min \left\| \begin{bmatrix} \bar{A} \\ \bar{B} \\ C \\ \lambda S \end{bmatrix} z^{n+1} - \begin{bmatrix} \mathbf{d} \\ \mathbf{q} \\ \mathbf{p} \\ \lambda k \end{bmatrix}^n \right\|_2 \tag{2.7}$$

Equation (2.7) is equivalent to

$$\overline{\mathbf{W}}z = \bar{\mathbf{v}} \tag{2.8}$$

where $\overline{\mathbf{W}} = \bar{\mathbf{A}}^T\bar{\mathbf{A}} + \bar{\mathbf{B}}^T\bar{\mathbf{B}} + \bar{\mathbf{C}}^T\bar{\mathbf{C}} + \lambda^2\mathbf{S}^T\mathbf{S}$ is a symmetric positive definite large sparse matrix and $\bar{\mathbf{v}} = \bar{A}^T\mathbf{d} + \bar{B}^T\mathbf{q} + \bar{C}^T\mathbf{p} + \lambda^2\mathbf{S}^T\mathbf{k}$.

Equation (2.8) is the algebraic system solved by HASM.MOD. Figure 2.1 presents the distribution of non-zero elements in $\overline{\mathbf{W}}$, where $I = J = 8$, $(I+1)\times(J+1) = 81$. The order of the coefficient matrix is 81×81, and nz is the number of non-zero elements.

Fig. 2.1 Distribution of non-zero elements in the HASM.MOD matrix

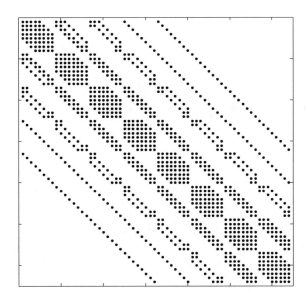

2.1.1 Numerical Simulation

In this study, the simulation accuracy of HASM before and after improvement is compared using eight mathematical surfaces. These surfaces are widely used in real life because of their special geometric properties. The analytic expressions of the surfaces and the corresponding study areas are shown in Fig. 2.2.

$$f_1 : f_1(x, y) = e^{\left\{-\frac{(5-10x)^2}{2}\right\}} + 0.75e^{\left\{-\frac{(5-10y)^2}{2}\right\}} + 0.75e^{\left\{-\frac{(5-10x)^2}{2}\right\}} + e^{\left\{-\frac{(5-10y)^2}{2}\right\}},$$ and the domain of definition is $\{(x, y)|0 \le x \le 1, 0 \le y \le 1\}$;

$f_2 : f_2(x, y) = sin(2\pi y) \cdot sin(\pi x)$, and the domain of definition is $\{(x, y)|0 \le x \le 1, 0 \le y \le 1\}$;

$$f_3 : \quad f_3 \quad = \quad 0.75e^{\left\{-\left[\frac{(9x-2)^2+(9y-2)^2}{4}\right]\right\}} + 0.75e^{\left[-\frac{(9x+1)^2}{49} - \frac{9y+1}{10}\right]},$$
$$+0.5^{\left\{-\left[\frac{(9x-7)^2+(9y-3)^3}{4}\right]\right\}} - 0.2e^{\left[-(9x-4)^2-(9y-7)^2\right]}$$, and the domain of definition is $\{(x, y)|0 \le x \le 1, 0 \le y \le 1\}$;

$f_4 : f_4(x, y) = \frac{1}{3}e^{-\left(\frac{81}{4}\right)\left[(x-0.5)^2+(y-0.5)^2\right]}$, and the domain of definition is $\{(x, y)|0 \le x \le 1, 0 \le y \le 1\}$;

$f_5 : f_5(x, y) = \frac{1}{3}e^{-\left(\frac{81}{16}\right)\left[(x-0.5)^2+(y-0.5)^2\right]}$, and the domain of definition is $\{(x, y)|0 \le x \le 1, 0 \le y \le 1\}$;

$f_6 : f_6(x, y) = 3(1-x)^2 e^{-x^2-(y+1)^2} - 10\left(\frac{x}{5} - x^3 - y^5\right)e^{-x^2-y^2} - \frac{e^{-(x+1)^2-y^2}}{3}$, and the domain of definition is $\{(x, y)| -3 \le x \le 3, -3 \le y \le 3\}$;

$f_7 : f_7(x, y) = x \cdot y$, and the domain of definition is $\{(x, y)|0 \le x \le 1, 0 \le y \le 1\}$;

$f_8 : f_8(x, y) = x^2 - y^2$, and the domain of definition is $\{(x, y)|x^2 + y^2 \le 1\}$.

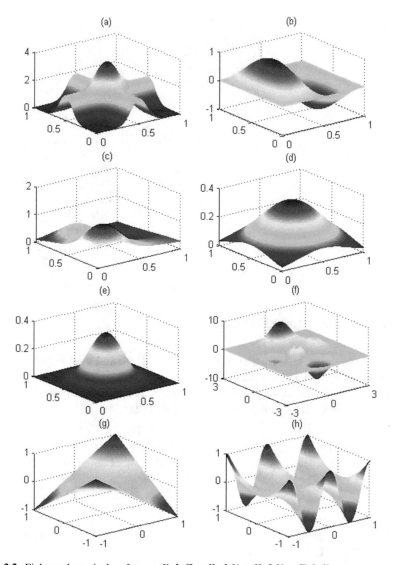

Fig. 2.2 Eight mathematical surfaces. **a** f1. **b** f2. **c** f3. **d** f4. **e** f5. **f** f6. **g** f7. **h** f8

To compare the simulation accuracy of traditional HASM with that of HASM.MOD when convergence is reached, the sampling ratio is set at 1/100, and the stopping criteria are $\|b - Ax\|_2 \leq 10^{-12}$ for the inner iteration and 5 outer iterations. Traditional HASM is often used only as a correction method for other interpolation methods; that is, the interpolation results of other methods are used as the iterative driving field in HASM to further correct the results obtained by other methods. Therefore, the simulation results of the inverse distance weight (IDW) are used as the driving fields of both HASM and HASM.MOD in this chapter. Table

Table 2.1 Calculation errors of HASM and HASM.MOD

Method	f_1	f_2	f_3	f_4	f_5	f_6	f_7	f_8
HASM.MOD	0.00580	0.00395	0.00165	0.0001	0.00032	0.0152	0.0003	0.0032
HASM	0.3270	0.0402	0.0140	0.0031	0.0028	0.2132	0.0089	0.0290

Table 2.2 Differences between simulation results of HASM and HASM.MOD and real values

Method		f_1	f_2	f_3	f_4	f_5	f_6	f_7	f_8
ΔMax	HASM.MOD	0.02	0	0.01	0.0001	0.003	0.05	0	0
	HASM	9.28	0.01	0.02	0.0001	0.003	0.58	0.05	0.02
ΔMin	HASM.MOD	0	0	0	0.0007	0	0.02	0.02	0
	HASM	2.00	0.01	0	0.0007	0	0.06	0	0.08
ΔMean	HASM.MOD	0	0	0.001	0.0006	0	0.0001	0	0
	HASM	0.01	0	0.0003	0.0006	0.0001	0.008	0	0.0005

2.1 shows the root mean square errors (RMSEs) of the performance of HASM and HASM.MOD, where $RMSE = \sqrt{\frac{1}{N} \sum_{k=1,...,N} (z_k - \overline{z_k})^2}$, in which z_k is the true value of the kth grid point (x_i, y_j), $\overline{z_k}$ is the corresponding simulated value, and N is the number of grid points.

Table 2.1 shows that the accuracy of HASM.MOD is greater than that of the traditional HASM method for all eight surfaces. Table 2.2 shows the differences between the maximum, minimum and mean simulated values of HASM and HASM.MOD and the real values.

Table 2.2 also shows that the performance of HASM.MOD is better than that of the traditional HASM method. In particular, for surfaces f_1 and f_6, the extreme maximum and minimum values occur in the traditional HASM. The results of HASM and HASM.MOD for Gaussian surface f_6 and saddle surface f_7 on a rectangular area are shown in Figs. 2.3 and 2.4, respectively.

The results of HASM significantly oscillate when the IDW interpolation results are used as the driving field.

The first fundamental coefficients E, F, and G of a surface reflect the implicit nature of the surface, such as the length of a curve on the surface, the area of a region on the surface, and the angle between two curves on the surface; and the second fundamental coefficients L, M, and N reflect the local curvature variation of the surface. To more accurately depict the simulation effects before and after HASM improvement, the six fundamental coefficients of the simulated surfaces of HASM and HASM.MOD are compared. Figure 2.5 (right) shows the six fundamental coefficients of the Gaussian surface simulated using the traditional HASM method, and Fig. 2.5 (left) shows the actual six fundamental coefficients. From top to bottom, they are E, F, G, L, M, and N. Figure 2.6 (right) shows the six fundamental coefficients of the Gaussian surface simulated using HASM.MOD, and Fig. 2.6

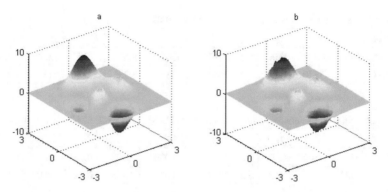

Fig. 2.3 Gaussian surface simulation. **a** HASM.MOD. **b** HASM

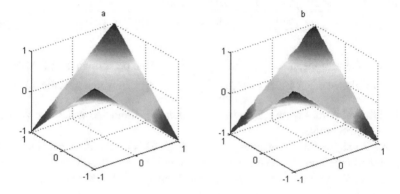

Fig. 2.4 Saddle surface simulation. **a** HASM.MOD. **b** HASM

(left) shows the actual six fundamental coefficients. From top to bottom, they are E, F, G, L, M, and N.

The local information of the simulated surface shows that HASM.MOD performs significantly better than the HASM method. In Fig. 2.5, extreme maximum and minimum values occur in the six fundamental coefficients simulated by HASM, and the values differ greatly from the real values. Moreover, HASM does not completely portray the second fundamental coefficients $L, M, and N$ that characterize the local details of the surface. In comparison, the accuracy of the six fundamental coefficients $E, F, G, L, M, and N$ simulated by HASM.MOD is higher than that by the HASM method.

Figures 2.7 and 2.8 compare the six simulated fundamental coefficients of the saddle surface on a rectangular area and show that the six fundamental coefficients of the saddle surface simulated by HASM.MOD are significantly better than those by the traditional HASM method. HASM does not completely portray either the first fundamental coefficients $E, F, and G$, which represent the implicit properties of the surface, or the second fundamental coefficients $L, M, and N$, which characterize

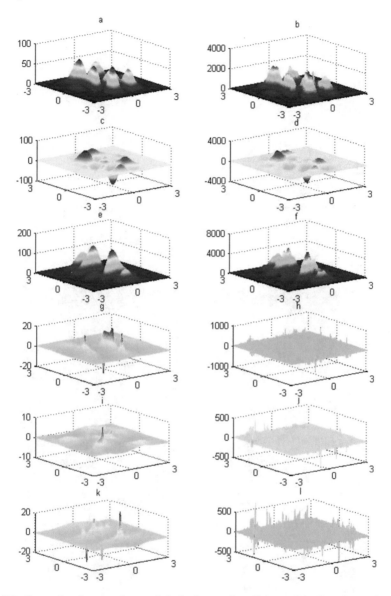

Fig. 2.5 Comparison between the actual six fundamental coefficients of the Gaussian surface and the HASM simulated values. **a** E_real. **b** E_simu. **c** F_real. **d** F_simu. **e** G_real. **f** G_simu. **g** L_real. **h** L_simu. **i** M_real. **j** M_simu. **k** N_real. **l** N_simu

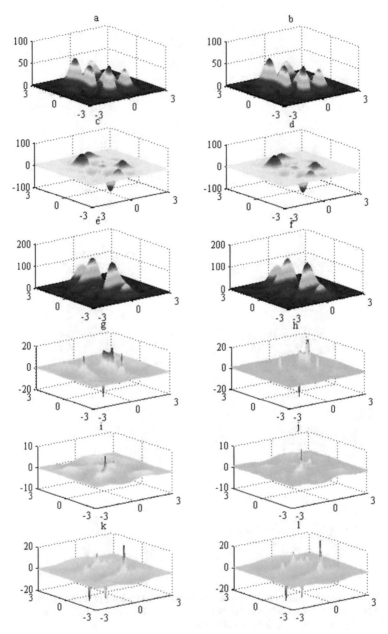

Fig. 2.6 Comparison between the actual six fundamental coefficients of the Gaussian surface and the HASM.MOD simulated values. **a** E_real. **b** E_simu. **c** F_real. **d** F_simu. **e** G_real. **f** G_simu. **g** L_real. **h** L_simu. **i** M_real. **j** M_simu. **k** N_real. **l** N_simu

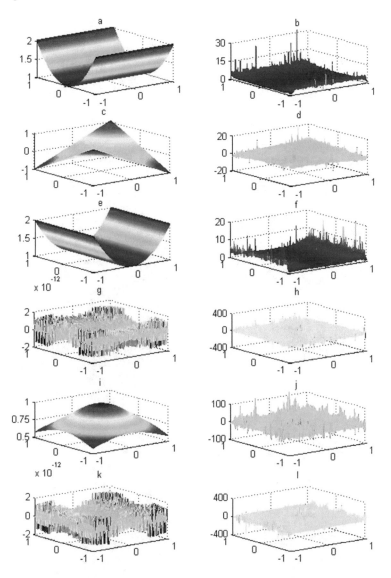

Fig. 2.7 Comparison between the actual six fundamental coefficients of the saddle surface and the HASM simulated values. **a** E_real. **b** E_simu. **c** F_real. **d** F_simu. **e** G_real. **f** G_simu. **g** L_real. **h** L_simu. **i** M_real. **j** M_simu. **k** N_real. **l** N_simu

the local details of the surface. As seen in Fig. 2.8 (left), the six fundamental coefficients simulated by HASM.MOD are very close to those of the real surface, and relatively large errors are observed for the simulated second fundamental coefficients L and N because the actual values of L and N are close to 0. Due to the influence of computational perturbations, rounding errors, and other factors, the differences

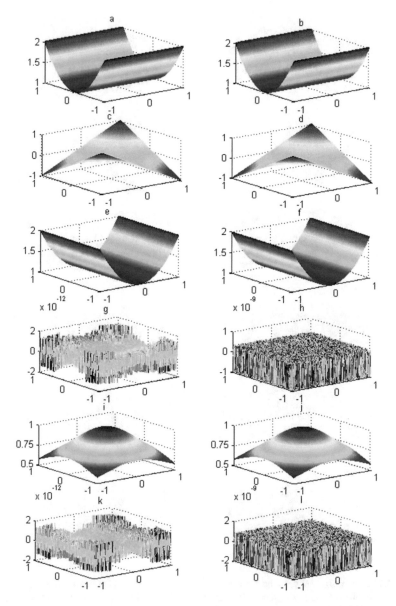

Fig. 2.8 Comparison between the actual six fundamental coefficients of the saddle surface and the HASM.MOD simulated values. **a** E_real. **b** E_simu. **c** F_real. **d** F_simu. **e** G_real. **f** G_simu. **g** L_real. **h** L_simu. **i** M_real. **j** M_simu. **k** N_real. **l** N_simu

between the simulated and real values of L and N are more obvious than those of other fundamental coefficients, with a magnitude of error of $O(10^{-8})$. For other fundamental coefficients with large values, the simulated value is not significantly different from the real one due to the small simulation error of HASM.MOD.

Thus, according to the simulation of the surface and the details of the surface, the simulation accuracy of HASM.MOD is significantly better than that of the traditional HASM method.

2.1.2 Case Study

Many real-life problems require data interpolation and the use of completely spatialized data for analysis. As the most important climate elements, temperature and precipitation are the key indicators for describing the climate changes in a region or even for the earth, and the changes in spatial distribution directly affect industrial and agricultural production, which in turn restrict economic development.

This book takes temperature and precipitation as examples to investigate the simulation performance of HASM.MOD to provide more accurate spatial distribution data for related studies.

2.1.2.1 Study Area

Located in the southeast of Eurasia and bordering the Pacific Ocean (15–55°N, 70–140°E), China has a vast area and complex topography. With the Tibetan Plateau in the west, the western Pacific Ocean in the east, complex topography and vegetation in the north, and the equatorial ocean in the south, the complexity and diversity of topography make the climate more complex and diverse. Generally, the northwest of China is a temperate continental climate zone, the Qinghai-Tibet Plateau has an alpine climate, and the southeast coast has an oceanic climate, while most areas are in a monsoon climate zone. From the perspective of temperature zones, there are tropical, subtropical, warm temperate, middle temperate, cold temperate, and Qinghai-Tibet Plateau zones in China. Due to the complex and diverse topography of China, over 60% of the area is covered by mountains and hills, and different scales of topography have significantly different effects on climate. The distribution of precipitation in China decreases from the southeast coastal area to the northwest inland area. The regional precipitation differences are large, with the pattern of coastal areas > inland areas, southern areas > northern areas, and mountainous areas > plain areas. Monsoon activities have a significant impact on precipitation in China. Precipitation in China exhibits obvious seasonality, with more precipitation in summer and autumn and less precipitation in winter and spring in most parts of the country.

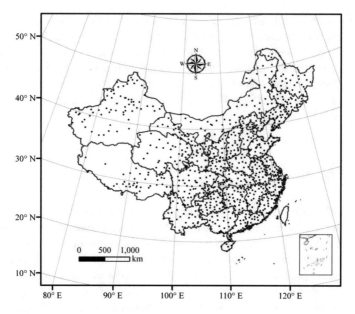

Fig. 2.9 Distribution of meteorological stations in China

2.1.2.2 Distribution of Meteorological Stations

The data used are monthly average temperature and precipitation data of 752 meteorological stations across China since 1951 and the longitude and latitude data of the stations. The meteorological stations are unevenly distributed in China, with a dense distribution in the eastern region, and compared to eastern region, the stations in the western region are extremely unevenly distributed and sparse. Generally, more stations are located in economically developed areas and fewer stations are located in undeveloped areas (Fig. 2.9). The stations are mainly located in plains and river valleys, and few stations are located on mountain slopes and hilltops. The distance between meteorological stations is approximately 100 km. In addition, the meteorological stations are not established at the same time. Before 1950, there were very few meteorological stations west of 100 °E in mainland China, and large-scale construction of stations began in the early 1950s.

2.1.2.3 Spatial Interpolation of Meteorological Elements

For the spatial interpolation of meteorological elements, the interpolation accuracy is improved during hybrid interpolation by correcting the residuals presented in the global interpolation. The hybrid interpolation combines the advantages of other interpolation methods and is an important research direction for future meteorological interpolation. In the HASM model, only the mathematical properties of the surface and the observation data at the sampling stations are utilized, without consideration of the factors affecting temperature and precipitation. For smooth areas (low-frequency oscillation), the trend surface analysis can be used for interpolation, while in areas with dramatic changes (high-frequency oscillation), the HASM method can be used (Yue et al. 2013a, b). The background values of the trend surface can effectively compensate for the limitations of surface modelling when the sampling points are sparse. Moreover, a good surface simulation method should be able to capture the local information missed by the spatial trend surface. Through various auxiliary variables associated with the predictor variables, HASM is able to reduce the uncertainty and improve the accuracy of the simulation. The estimation of the trend surface is generally calculated using multiple regression or polynomial regression based on the least-squares principle. Previous studies (Burrough 1996; Yan et al. 2005) have shown that the more explanatory variables introduced in the trend surface analysis, the more accurate the results are. However, introducing more variables can cause a rapid increase in the computation time. Thus, the principle is usually to, without compromising the accuracy of the results, first use a simple method to estimate the background field. Once the trend is determined, the remaining part is iteratively corrected using HASM to improve the accuracy based on the principal theorem of surface theory.

High Accuracy Surface Simulation of Temperature

China's terrain is very diverse, and the differences between the east and west regions are large. There are numerous factors that affect the spatial distribution of temperature in China, among which the altitude and latitude are the most significant. In general, the temperature declines with increasing altitude. The influence of altitude error on the accuracy of the spatial interpolation of temperature cannot be ignored. In view of this, some scholars propose that a 0.6 °C decrease should be used for every 100 m increase in altitude to normalize the temperature according to the temperature at sea level (Song et al. 2006). However, the direct temperature reduction rate varies considerably with time and space, and the above method may introduce large errors. Since a digital elevation model (DEM) raster file is required to generate the trend surface raster map, the altitude information extracted from the DEM raster file is used as one of the explanatory variables for temperature in this chapter to reduce the errors. The latitude is another significant factor that affects the temperature. Since places with different latitudes have different solar altitude angles, the temperature varies greatly from place to place. Generally, the lower the latitude is, the higher the temperature, and the higher the latitude is, the lower the temperature. The spatial distribution of the temperature in China is greatly influenced by the topography, and it is necessary to consider the influence of latitude on temperature when using limited observation points to simulate a reasonable spatial distribution of temperature under complex topographical conditions. For simplicity, only altitude and latitude are considered in the trend surface analysis in this chapter. Temperature simulation can be carried out by using a linear model of ordinary least squares (OLS) combined

Table 2.3 Simulation accuracies of HASM and HASM.MOD for monthly average temperature

Method	HASM		HASM.MOD	
Month	January	July	January	July
MAE (°C)	1.22	0.94	1.11	0.85
RMSE (°C)	1.73	1.42	1.59	1.26

with residual HASM iterative simulation to fit the temperature field:

$$Y_{TEM} = X^T \theta_{OLS} + X_{HASM},$$

where $X = [I, DEM, Lat]^T$ and I is a column vector with all elements being 1 and its length being the number of sampling points. DEM is the altitude (in m), Lat is the latitude of the station (in m), $\theta_{OLS} = (\mathbf{X}^T\mathbf{X})^{-1}\mathbf{X}^T y$ is the coefficient of each explanatory variable calculated by OLS, y is the multi-year monthly average temperature at the sampling point, and $\mathbf{X}_{HASM}$ is the residual correction value obtained by HASM after removing the trend, in which $X_{HASM} \approx A^{-1}b$. $\mathbf{A}$ and $\mathbf{b}$ are the coefficient matrix and the right end term of the set of equations of the HASM model, respectively, and Y_{TEM} is the final simulated temperature.

The above method is used to simulate the average temperature in January and July from 1951 to 2010, and the spatial resolution is 10 km. The accuracies of HASM and HASM.MOD are compared.

To calculate the average temperature in January, a linear fit based on OLS is used to obtain the temperature trend surface equation:

$$X^T \theta_{OLS} = [I, \ DEM, Lat][41.1909, 0.0027, 0.000011]^T$$

Similarly, the expression of the trend surface of the average temperature in July is:

$$X^T \theta_{OLS} = [I, DEM, Lat][33.1623, 0.0040, 0.000016]^T$$

The residual X_{HASM} after removing the trend is interpolated using HASM and HASM.MOD. Then, the result Y_{TEM} is obtained.

In the study area, 15% of the points are selected for validation, and 85% of the points are selected for simulation. The simulation accuracies of HASM and HASM.MOD are shown in Table 2.3.

The mean absolute error MAE $= \frac{1}{N}\sum_{i=1}^{N}(y_{TEM} - y)$, RMSE $= \sqrt{\frac{1}{N}\sum_{i=1}^{N}(y_{TEM} - y)^2}$, and N is the number of validation points. Table 2.3 shows that the accuracy of HASM.MOD is better than that of HASM for both January and July temperatures.

Figures 2.10 and 2.11 show that the differences between HASM.MOD and HASM are more pronounced in a few regions. Specifically, in January, the differences in

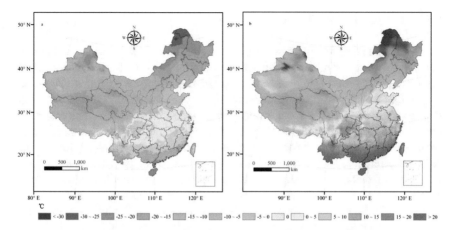

Fig. 2.10 Simulation of average temperature in January. **a** HASM. **b** HASM.MOD

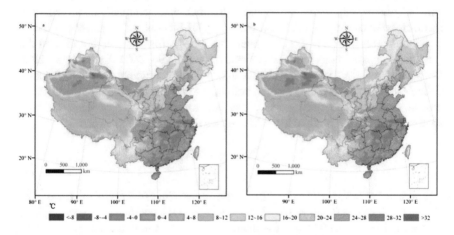

Fig. 2.11 Simulation of average temperature in July. **a** HASM. **b** HASM.MOD

Xinjiang, Heilongjiang, Inner Mongolia, and Anhui are significant, and in July, the difference in Xinjiang is significant. In this study, the spatial distribution field of temperature is decomposed into a smooth part and a strongly varying part, which are fitted by least-squares regression and HASM interpolation, respectively. The accuracy of the trend surface plays a crucial role in the interpolation results of the hybrid interpolation. For the simulation of temperature in January and July, there are no significant differences between the two methods in most parts of China. Nonetheless, Table 2.3 indicates that the overall simulation accuracy of HASM.MOD is better than that of HASM.

High Accuracy Surface Simulation of Precipitation

Relatively speaking, temperature variation is small over a large area, and the relationship between the independent and dependent variables is assumed to be constant across the study area using OLS; i.e., spatial stationarity is assumed (Yue et al. 2013a, b). OLS regression is a global model that cannot handle local spatial heterogeneity, and one consequence of using global models is that spatial data errors and uncertainties tend to aggregate. Compared to temperature, precipitation has strong local characteristics, and the observation data have large uncertainties, which results in great differences in precipitation among different methods. Among the various geographic factors, topography and altitude have the most significant influence on precipitation. As the airflow is lifted along the slope, precipitation generally increases with the altitude. However, there might be anomalies due to the influence of topography and other factors. For instance, the precipitation above 1400 m in the Qinling Mountains decreases with altitude. In China, precipitation varies with latitude due to atmospheric circulation and movement of monsoon rain bands. Since there are complex relationships between the spatial distribution of precipitation and geographical factors, obtaining a global relationship by OLS is not suitable. Geographically weighted regression (GWR) is a nonparametric local linear regression method that can address spatial heterogeneity (Brunsdon et al. 1996, 2001; Leung et al. 2000; Dormann et al. 2007; Wang et al. 2012). In GWR, the coefficient of each explanatory variable varies with the study area; therefore, the variable coefficients are critical for reflecting the nonstationary spatial characteristics. In this chapter, a nonstationary spatial-based HASM method is used to reduce the surface trend errors:

$$Y_{PRE} = X^T \theta_{GWR} + X_{HASM}$$

where $X = [I, DEM, Slope]^T$; DEM is the altitude (in m); Slope is the slope of the station (in degrees); $\theta_{GWR} = (X^T W X)^{-1} X^T W y$ is the coefficient of each explanatory variable in the weighted least-squares method, which varies with the sampling location; and y is the multi-year monthly average precipitation at the sampling point. X_{HASM} is the residual correction value using the HASM method after removing the trend, and $X_{HASM} \approx A^{-1} b$. A and **b** are the coefficient matrix and the right end term of the set of equations of the HASM model, respectively, and Y_{PRE} is the simulated precipitation.

The multi-year average precipitation distribution in January and July are simulated based on GWR and HASM residual interpolation, and the results are shown in Table 2.4.

Figures 2.12 and 2.13 show the simulated distribution of precipitation in January and July using HASM and HASM.MOD, and there are no significant differences between the two models, with consistent distribution trends. However, in some areas, such as central Shandong in January, the simulated value of HASM is lower than the real value.

Table 2.4 Simulation of monthly average precipitation

Method	HASM		HASM.MOD	
Month	January	July	January	July
MAE (mm)	2.19	21.93	1.90	20.92
RMSE (mm)	2.75	29.79	2.56	26.59

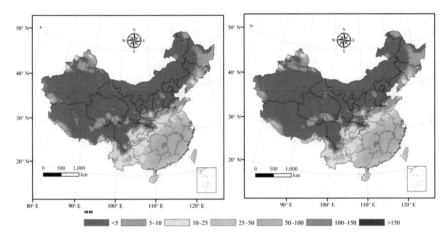

Fig. 2.12 Simulation of the average precipitation in January. **a** HASM. **b** HASM.MOD

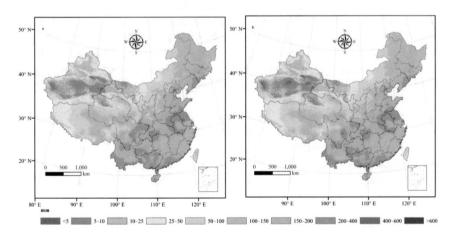

Fig. 2.13 Simulation of the average precipitation in July. **a** HASM. **b** HASM.MOD

For the simulation of temperature and precipitation, a hybrid interpolation method is used, where HASM is used as a correction method for the residuals after removing the trend. According to the simulation results, the overall difference between HASM

and HASM.MOD is not significant, yet the accuracy of HASM.MOD is higher than
that of HASM.

2.2   Reasons for Accuracy Improvement

In this chapter, the discretization of Eq. (2.2) in traditional HASM is performed by
the high-order difference scheme, and the effect of the high-order difference scheme
on the accuracy of HASM simulation is investigated; additionally, traditional HASM
using the high-order difference scheme is compared with HASM.MOD to analyse
the factors in the accuracy improvement of the HASM model.

The traditional HASM with the high-order difference scheme using the Gaussian
surface is denoted HASM1, and the partial derivatives in Eq. (2.2) are Eqs. (2.3) and
(2.4). The modern HASM with the low-order difference scheme is denoted HASM2,
and the difference scheme of f_x, f_{xx}, f_y, and f_{yy} is Eq. (1.6).

Table 2.5 shows the results when the number of grids is different. The stopping
criterion for the inner iteration is $\left\| \mathbf{b}^{(k)} - \mathbf{A}\mathbf{x}^{(k+1)} \right\|_2 \leq 10^{-12}$, and the number of
outer iterations is 5.

The simulation results show that HASM1, with high-accuracy difference scheme,
yields better results than HASM, yet the error increases for both models with the
number of grids when the same number of outer iterations is used. The results of
HASM and HASM2 show that the simulation accuracy of HASM is significantly
improved after including the third equation, and its accuracy is better than that of
HASM1, which considers only the with high-accuracy difference scheme. Further,
HASM.MOD demonstrates the highest accuracy, with the addition of the third equation and the use of the high-accuracy difference scheme. Moreover, as the number of
grids increases, the errors of HASM2 and HASM.MOD do not increase significantly
at the given number of outer iterations, indicating that the models considering all
partial differential equations are more stable and accurate.

The analysis results show that the main reason for the improved simulation accuracy is the introduction of the third equation in the Gaussian equations (the equation
satisfied by the mixed partial derivatives) and the high-accuracy difference scheme,
and the effect of the high-accuracy difference scheme is weaker than that of the
third equation. Thus, a high-accuracy difference scheme does not necessarily give

Table 2.5 Computational errors (RMSE) of different discretization schemes of the three Gaussian
equations using different numbers of grids

Number of grids	289	625	1296	2209	3721
HASM	0.2021	0.2543	0.3109	0.3593	0.3958
HASM1	0.1959	0.2500	0.2935	0.3393	0.3485
HASM2	0.0328	0.0436	0.0868	0.0995	0.0946
HASM.MOD	0.0266	0.0423	0.0719	0.0880	0.0890

a good approximate solution, and a reasonable difference scheme needs to preserve some physical properties of the original solution. The introduction of the mixed partial derivative term in HASM.MOD helps to describe detailed local features of the surface.

2.3 Boundary Value Problem

Since the sampling points around an estimation point at the boundary are sparse and there are no sampling points outside the boundary, the simulation accuracy of HASM at the boundary is low. Although the spatial error is solved by Taylor series expansion, which increases the accuracy of HASM, the improvement is not obvious. Studies have shown that boundary data with a high accuracy can help improve the overall interpolation accuracy of HASM (Chen et al. 2009). Traditionally, the simulation accuracy at the boundary relies on other interpolation methods or Laplace's equation. Chen et al. (2009) reported that the Laplace equation was more effective than other methods in simulating the boundaries; however, the method does not have a theoretical basis. Moreover, the boundary data are not involved in the HASM iterations during the simulation process, which further limits the accuracy at the boundary. In this chapter, HASM.MOD does not separately consider the boundary values, yet the boundary and the inner region are treated equally and the boundary values are the iterative driving fields of HASM.MOD. The boundary values are used in the iterative correction process during the solution of HASM.MOD. Figure 2.14 shows the simulation areas of HASM and HASM.MOD, in which the blue box is the simulation area of HASM.

The finite-difference scheme at the boundary in HASM.MOD is:

$$(f_x)_{(i,j)} \approx \begin{cases} \frac{-3f_{0,j}+4f_{1,j}-f_{2,j}}{2h} & i=0, j=0,\ldots,J+1 \\ \frac{f_{i+1,j}-f_{i-1,j}}{2h} & i=1,\ldots,I, j=0,J+1 \\ \frac{3f_{I+1,j}-4f_{I,j}+f_{I-1,j}}{2h} & i=I+1, j=0,\ldots,J+1 \end{cases},$$

$$(f_{xx})_{(i,j)} \approx \begin{cases} \frac{2f_{0,j}-5f_{1,j}+4f_{2,j}-f_{3,j}}{h^2} & i=0, j=0,\ldots,J+1 \\ \frac{-f_{i+2,j}+16f_{i+1,j}-30f_{i,j}+16f_{i-1,j}-f_{i-2,j}}{12h^2} & i=1,\ldots,I, j=0,J+1, \\ \frac{2f_{I+1,j}-5f_{I,j}+4f_{I-1,j}-f_{I-2,j}}{h^2} & i=I+1, j=0,\ldots,J+1 \end{cases}$$

$$(f_y)_{(i,j)} \approx \begin{cases} \frac{-3f_{i,0}+4f_{i,1}-f_{i,2}}{2h} & i=0,\ldots I+1, j=0 \\ \frac{f_{i,j+1}-f_{i,j-1}}{2h} & i=0, I+1, j=1,\ldots J, \\ \frac{3f_{i,J+1}-4f_{i,J}+f_{i,J-1}}{2h} & i=0,\ldots I+1, j=J+1 \end{cases}$$

$$(f_{yy})_{(i,j)} \approx \begin{cases} \frac{2f_{i,0}-5f_{i,1}+4f_{i,2}-f_{i,3}}{h^2} & i=0,\ldots,I+1, j=0 \\ \frac{-f_{i,j+2}+16f_{i,j+1}-30f_{i,j}+16f_{i,j-1}-f_{i,j-2}}{12h^2} & i=0, I+1, j=1,\ldots,J, \\ \frac{2f_{i,J+1}-5f_{i,J}+4f_{i,J-1}-f_{i,J-2}}{h^2} & i=0,\ldots I+1, j=J+1 \end{cases}$$

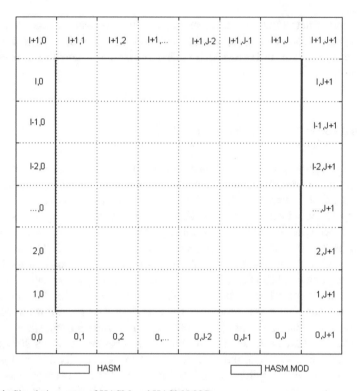

Fig. 2.14 Simulation areas of HASM and HASM.MOD

$$(f_{xy})_{(i,j)} \approx \begin{cases} \frac{f_{1,1}-f_{1,0}-f_{0,1}+f_{0,0}}{h^2} & i=0, j=0 \\ \frac{f_{1,J+1}+f_{0,J}-f_{1,J}-f_{0,J+1}}{h^2} & i=0, j=J+1 \\ \frac{f_{1,j+1}-f_{0,j+1}+f_{0,j-1}-f_{1,j-1}}{2h^2} & i=0, j=1,\cdots J \\ \frac{f_{I+1,1}-f_{I,0}-f_{I,1}+f_{I+1,0}}{h^2} & i=I+1, j=0 \\ \frac{f_{I,J}-f_{I+1,J}-f_{I,J+1}+f_{I+1,J+1}}{h^2} & i=I+1, j=J+1 \\ \frac{f_{I+1,j+1}-f_{I,j+1}+f_{I,j-1}-f_{I+1,j-1}}{2h^2} & i=I+1, J=1,\ldots,J \\ \frac{f_{i+1,1}-f_{i+1,0}+f_{i-1,0}-f_{i-1,1}}{2h^2} & i=1,\cdots I, j=0 \\ \frac{f_{i+1,J+1}-f_{i+1,J}+f_{i-1,J}-f_{i-1,J+1}}{2h^2} & i=1,\cdots, I, j=J+1 \end{cases}$$

After the analysis of the main causes of HASM errors, the HASM itself is further studied to reduce the errors at the boundary by taking the Gaussian surface as an example. The simulation accuracies of HASM and HASM.MOD at the outermost, outer and second outer layers are studied, and the simulation effects at the four sides (top, bottom, left, and right) are compared.

Numerical studies have shown that after several iterations, HASM can ensure that the simulation values in the calculation region do not oscillate (Yue and Du 2005). However, after several simulations, the calculation accuracy of HASM at the

Table 2.6 Computational errors of HASM and HASM.MOD at the boundaries (number of grids: 441)

HASM	0		1		2	
	Top	0.0497	Top	0.6178	Top	1.0765
	Bottom	0.0401	Bottom	0.7506	Bottom	1.3122
	Left	0.0063	Left	0.2125	Left	0.3256
	Right	0.0055	Right	0.2375	Right	0.4217
HASM.MOD	0		1		2	
	Top	0.0163	Top	0.3525	Top	0.9053
	Bottom	0.0392	Bottom	0.4366	Bottom	1.1186
	Left	0.0165	Left	0.0596	Left	0.1818
	Right	0.0166	Right	0.0597	Right	0.1927

boundary is low, and the calculation error at the boundary has a large impact on the simulation results of the whole region (Yue 2011). Table 2.6 presents the computational errors at the boundaries when HASM and HASM.MOD reach convergence. The outermost layer is labelled "0", and the label of the layer increases along the inward direction. The RMSEs of HASM and HASM.MOD at the top, bottom, left, and right boundaries are shown in Table 2.6.

Since the result of the outermost layer is obtained using other interpolation methods, the calculation area in HASM is the inner area without the outermost layer. The error of HASM at the outermost boundaries is relatively low, and the error at layers 1 and 2 is high. In comparison, the computational errors of HASM.MOD at different layers are quite close. The errors of HASM.MOD are lower than those of HASM at layers 1 and 2. There are large gaps in the errors between layers 0 and 1; thus, traditional HASM exhibits the problem of an oscillating boundary. In HASM.MOD, the problem of oscillating boundary is eliminated to a certain extent.

2.4 Stopping Criteria for Iteration in HASM

According to the principal theorem of surface theory, a surface can be uniquely determined by Gaussian equations when the first and second fundamental coefficients of the surface satisfy the Gauss-Codazzi equations. Moreover, the whole outer iteration process of HASM corrects the right end term so that it gradually satisfies the Gauss-Codazzi equations. In HASM.MOD, finite-difference discretization of the Gaussian equations is performed to transform the problem into the following algebraic equation:

$$\mathbf{W}z^{(k+1)} = \mathbf{v}^{(k)}, \tag{2.9}$$

where $\mathbf{W} \in \mathbf{R}^{n \times n}$ is the n order-symmetric positive definite sparse matrix, $z \in \mathbf{R}^n$, k is the number of outer iterations, and the outer iterations are the process of updating $\mathbf{v}$. The process of solving Eq. (2.9) is performed by the inner iterations. For a given practical problem, Eq. (2.9) to be solved by HASM.MOD is fixed; thus, the stopping criterion for the inner iteration can be set as the iterative convergence criterion of the algebraic equation. According to the characteristics of the coefficient matrix in Eq. (2.9), the conjugate gradient (CG) method can be used to solve this equation, and the stopping criterion is $\|r\|_2 \le 10^{-12}$, where $r = \mathbf{v}^{(k)} - \mathbf{W}z^{(k+1)}$.

In traditional HASM, a set number of outer iterations (e.g., 5) is generally used. However, the first and second fundamental coefficients of the surface may not satisfy the Gauss-Codazzi equation with the predefined number of outer iterations. In view of this problem, based on the principal theorem of surface theory, the stopping criterion for iteration in HASM.MOD, that is, the determination of whether the Gauss-Codazzi equation is satisfied, is proposed. This method enhances the theoretical basis of the model. Using the Gauss-Codazzi equation as the convergence stopping criterion helps unify the convergence criteria of HASM models, and a quantitative iteration stopping indicator can be defined for practical problems of different scales.

The Gauss-Codazzi equation is expressed as follows:

$$
\begin{cases}
\left(\frac{L}{\sqrt{E}} \right)_y - \left(\frac{M}{\sqrt{E}} \right)_x - N \frac{\sqrt{E}_y}{G} - M \frac{\sqrt{G}_x}{\sqrt{EG}} = 0 \\
\left(\frac{N}{\sqrt{G}} \right)_x - \left(\frac{M}{\sqrt{G}} \right)_y - L \frac{\sqrt{G}_x}{E} - M \frac{\sqrt{E}_y}{\sqrt{EG}} = 0 \\
\left(\frac{\sqrt{E}_y}{\sqrt{G}} \right)_y + \left(\frac{\sqrt{G}_x}{\sqrt{E}} \right)_x + \frac{LN - M^2}{\sqrt{EG}} = 0
\end{cases}
\tag{2.10}
$$

Assuming that $\frac{L}{\sqrt{E}} = \varphi_1$, $\frac{N}{\sqrt{G}} = \varphi_2$, $\frac{\sqrt{E}_y}{\sqrt{G}} = P$, $\frac{\sqrt{G}_x}{\sqrt{E}} = Q$, $\frac{M}{\sqrt{G}} = \phi_1$, $\frac{M}{\sqrt{E}} = \phi_2$, then the Gauss-Codazzi equation can be transformed into:

$$
\begin{cases}
\varphi_{1y} - \phi_{2x} - \varphi_2 P - \phi_1 Q = 0 \\
\varphi_{2x} - \phi_{1y} - \varphi_1 Q - \phi_2 P = 0 \\
Q_x + P_y + \varphi_1 \varphi_2 - \phi_1 \phi_2 = 0
\end{cases}
\tag{2.11}
$$

According to the principal theorem of surface theory, the stopping criterion for the outer iteration is to determine whether Eq. (2.11) is satisfied in the HASM model. Due to errors introduced by the finite-difference discretization of the partial differential terms in Eq. (2.11), the stopping criterion for outer iteration can be set as:

$$
\left(\varphi_{1y} - \phi_{2x} - \varphi_2 P - \phi_1 Q \right)^2 + \left(\varphi_{2x} - \phi_{1y} - \varphi_1 Q - \phi_2 P \right)^2
$$
$$
+ \left(Q_x + P_y + \varphi_1 \varphi_2 - \phi_1 \phi_2 \right)^2 <_\varepsilon .
\tag{2.12}
$$

where ε is the user-specified accuracy. During practical implementation, finite-difference discretization is performed on the partial differential terms in Eq. (2.12), and second-order central-difference discretization is used for the first-order partial

Table 2.7 Number of outer iterations upon convergence of HASM.MOD

Number of grids	289	625	1296	2209	3721
Number of outer iterations	112	360	867	1935	3622

Table 2.8 Errors (RMSE) of HASM.MOD under different convergence criteria for outer iteration

Number of grids	289	625	1296	2209	3721
Driving field error (kriging driving field)	0.2499	0.2808	0.3006	0.3068	0.3195
Five iterations	0.0266	0.0423	0.0719	0.0880	0.0890
Gauss-Codazzi	0.0020	0.0024	0.0070	0.0067	0.0061

derivatives:

$$f_x = \frac{f_{i+1,j} - f_{i-1,j}}{2h}, \quad f_y = \frac{f_{i,j+1} - f_{i,j-1}}{2h}$$

Through the introduction of the Gauss-Codazzi equation into the HASM.MOD model based on the principal theorem of surface theory, the theoretical basis of the model is enhanced. Moreover, the quantitative standards of the stopping criteria for inner and outer iterations in HASM.MOD can facilitate its further application.

Table 2.7 shows the number of outer iterations of HASM.MOD when the Gauss-Codazzi equation is satisfied at different computational scales when the Gaussian surface is taken as an example. The convergence criterion for the inner iteration is $\left\| \mathbf{b}^{(k)} - \mathbf{A}\mathbf{x}^{(k+1)} \right\|_2 \leq 10^{-12}$.

Table 2.7 shows that at different computational scales, the number of outer iterations needed to reach convergence is not constant and is significantly more than 5, as used in HASM. Therefore, the practice of taking a fixed number of outer iterations is not desirable and lacks a theoretical basis. Table 2.8 gives the same results in terms of iteration accuracy.

Table 2.8 shows that HASM.MOD is far from convergence at the given number of outer iterations.

The left end of Eq. (2.12) is denoted GC. Figure 2.15 shows the variation in GC with iteration when the number of grids is 2209. Figure 2.15a shows the variation during the first 100 iterations, and Fig. 2.15b shows the variation during the whole process. When the number of outer iterations is 5, the GC changes drastically and is much higher than 0. Thus, according to the principal theorem of surface theory, a surface cannot be obtained when the number of outer iterations is 5.

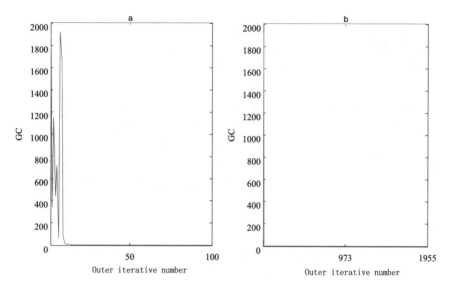

Fig. 2.15   Variation in GC (the left end of the Gauss-Codazzi equation; the number of grids is 2209)

2.5   Summary

In this chapter, traditional HASM is improved by introducing the equation satisfied by the mixed partial derivatives of the surface. Through numerical simulation and experimental verification, the superiority of HASM.MOD regarding the accuracy is validated. Moreover, the influence of the high-accuracy differential scheme on the simulation results is investigated, the simulation accuracy of HASM.MOD at the boundary is studied, and a quantitative stopping criterion for iteration is proposed. The results show the following:

(1)   Under the same conditions, the simulation accuracy of HASM.MOD is higher than that of traditional HASM. For mathematical surfaces, the differences between the simulation results of HASM and HASM.MOD are compared; additionally, the differences between the first and second fundamental coefficients of the surfaces of HASM and HASM.MOD are compared. The results of numerical simulation demonstrate the advantages of HASM.MOD regarding the simulation accuracy. In a case study, a combination of trend surface analysis with HASM residual interpolation is used, and the advantages of HASM.MOD over traditional HASM are not obvious. However, due to the introduction of the second fundamental coefficient M, which characterizes the local features of the surface, HASM.MOD is better than HASM at portraying the local details of surfaces.

(2)   The study of the effect of the high-order difference scheme on the accuracy of traditional HASM shows that the overall simulation accuracy is improved to some extent but that the improvement is not significant. In contrast, due to

the introduction of the equation satisfied by the mixed partial derivatives of the surface, HASM.MOD has significantly better accuracy than the traditional HASM method.

(3) Compared with traditional HASM, HASM.MOD performs iterative correction at the boundary of the simulated region. Numerical results show that the simulation accuracy of HASM.MOD at the boundary is better than that of the traditional HASM and that it eliminates the problem of an oscillating boundary to a certain extent.

(4) Based on the principal theorem of surface theory, a quantitative stopping criterion for iteration is proposed for HASM.MOD, and experiments show the effectiveness of the criterion, which promotes the intelligent development of HASM.

Note that HASM.MOD does not offer advantages in storage and computational speed over traditional HASM because of the added nonlinear equation satisfied by the mixed partial derivatives, which increases the computation complexity of HASM.MOD. The storage of HASM.MOD is approximately twice that of traditional HASM, and its computation time is approximately eight times that of traditional HASM. Considering the development of high-performance computers, high-accuracy spatial interpolation is more important than storage and computation speed.

References

Brunsdon C, Fotheringham AS, Charlton ME. 1996. Geographically weighted regression: A method for exploring spatial nonstationarity. Geographical Analysis, 28(4): 281–298.

Brunsdon C, Mcclatchey J, Unwin DJ. 2001. Spatial variations in the average rainfall-altitude relationship in Great Britian: An approach using geographical weighted regression. International Journal of Climatology, 21: 455–466.

Burrough PA. 1996. Principles of geographical information systems for land resources assessment. Oxford: Oxford University Press.

Chen, CF, Yue TX, Lu YM. 2009. Solutions for boundary errors in high precision surface modeling applications. Journal of Remote Sensing, 13: 458–462.

Dormann CF, McPherson MJ, Araujo MB, et al. 2007. Methods to account for spatial autocorrelation in the analysis of species distributional data: A review. Ecography, 30: 609–628.

Karniadakis GEM, Kirby II RM. 2003. Parallel scientific computing in C++ and MPI. Cambridge: Cambridge University Press.

Leung Y, Mei CL, Zhang WX. 2000. Statistical tests for spatial nonstationarity based on the geographically weighted regression model. Environment and Planning A, 32: 9–32.

Somasundaram D. 2005. Differential geometry. Harrow, U.K.: Alpha Science

Song YH, Ma JH, Liu F. 2006. Spatial distribution and regionalization of temperature in China based on GIS. Journal of Arid Land Resources and Environment, 4: 17–22.

Su BQ, Hu HS. 1979. Differential Geometry. Beijing: People's Education Press.

Wang CH, Zhang JS, Yan XD. 2012. The use of geographically weighted regression for the relationship among extreme climate indices in China. Mathematical Problems in Engineering, doi: https://doi.org/10.1155/2012/369539.

Yan H, Nix HA, Hutchinson MF, et al. 2005. Spatial interpolation of monthly mean climate data for China. International Journal of Climatology, 25: 1369–1379.

Yue TX, Du ZP. 2005. High precision surface modeling: the core module of new generation GIS and CAD.Progress in Natural Science, 15(3): 73–82.

Yue TX, Zhao N, Ramsey RD, et al. 2013. Climate change trend in China, with improved accuracy. Climatic Change, 120: 137–151.

Yue TX, Zhao N, Yang H, et al. 2013. A multi-grid method of high accuracy surface modeling and its validation. Transctions in GIS,17(6): 943–952.

Yue TX. 2011. Surface modeling: high accuracy and high speed methods. CRC Press.

Chapter 3
Sensitivity of HASM to the Selection of the Driving Field

The traditional high-accuracy surface modelling (HASM) method is used only as a correction method for other interpolation methods; in other words, the interpolation results of other methods are used as the driving field of HASM to further correct these results. Thus, traditional HASM does not function as a stand-alone surface modelling method. The modern HASM method (HASM.MOD) is built upon the fundamental theorem of surface theory, which introduces the equation satisfied by the mixed partial derivatives of the surfaces. In this chapter, the sensitivity of HASM and HASM.MOD to the selection of driving field is studied using eight mathematical surfaces and temperature and precipitation data.

3.1 Sensitivity of Traditional HASM to the Selection of the Driving Field

In traditional HASM, the driving field is calculated using other interpolation methods based on sampling data. Thus, the driving field obtained by different interpolation methods can have a significant influence on the results of HASM.

Taking the Gaussian surface as an example, the performance of traditional HASM with different driving fields (i.e., kriging interpolation, inverse distance weight (IDW) interpolation, spline interpolation, and zero driving field) is investigated. Table 3.1 shows the differences between the simulation results of traditional HASM with different driving fields and the real surface.

HASM_K refers to the kriging interpolation results as the driving field, HASM_I refers to the IDW interpolation results as the driving field, HASM_S refers to the spline interpolation results as the driving field, and HASM_0 uses zero driving field. Table 3.1 shows the maximum, minimum, mean, and standard deviation of the differences between the different methods and the real surface. The performance of traditional HASM differs greatly with different driving fields even when the same stopping criterion is used. The accuracy of HASM_0 is lower than that of HASM_S, while

Table 3.1 Simulation error of traditional HASM for the Gaussian surface with different driving fields

Method	HASM_K	HASM_I	HASM_S	HASM_0
Max	0.6056	1.7613	0.1973	0.2546
Min	0.6255	2.7547	0.1823	0.3346
Mean	0.0007	0.0076	0.0003	0.0011
Std	0.0663	0.2132	0.0161	0.0251

HASM_I produces the largest simulation error. Hence, although traditional HASM is used to iteratively correct the results of other interpolation methods, the simulation results are significantly influenced by the selected driving field. Moreover, in practical application, different driving fields should be explored to optimize the simulation accuracy of traditional HASM.

Figure 3.1 shows the simulation of the Gaussian surface using traditional HASM with different driving fields. The largest error is produced when IDW interpolation results are used as the driving field.

To further evaluate the performance of traditional HASM with different driving fields, Table 3.2 shows the differences between the real saddle surface and the results of traditional HASM with different driving fields.

For the simulation of the saddle surface, when zero driving field is used, the results of HASM_0 are significantly different from the real values, and the accuracy

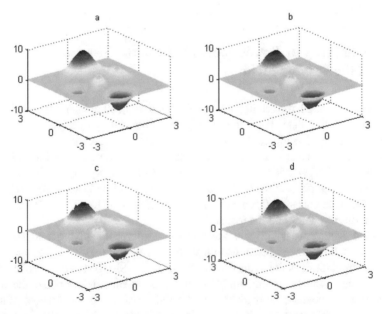

Fig. 3.1 Simulation of the Gaussian surface using different driving fields: **a** HASM_0, **b** HASM_K, **c** HASM_I, **d** HASM_S

Table 3.2 Simulation error of traditional HASM for the saddle surface with different driving fields

Method	HASM_K	HASM_I	HASM_S	HASM_0
Max	0.0585	0.105	0.0028	1
Min	0.0676	0.1200	0.0027	1.1243
Mean	7.62e−05	0.0004	0.492e−05	0.0006
Std	0.0022	0.0089	0.0002	0.1052

of HASM_0 is the lowest. Figure 3.2 shows the simulation of saddle surface using traditional HASM with different driving fields. HASM_0 can hardly restore the original surface, and the results of HASM_I are the second worst.

In summary, although traditional HASM uses the results of other interpolation methods as the driving field, the simulation accuracy strongly depends on the driving field. For different problems, the influence of the driving field on the simulation results of HASM is different. With zero driving field, the simulation accuracy of traditional HASM is poor. Therefore, in practice, it is necessary to test different driving fields to increase the accuracy of traditional HASM, which, of course, increases the complexity of the method and reduces its practicability. For users, the need for a universal simple driver field to simplify HASM is an urgent problem to be solved.

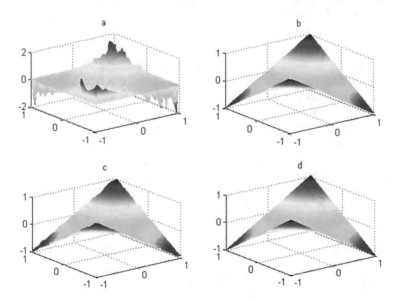

Fig. 3.2 Simulation of the saddle surface using different driving fields: **a** HASM_0, **b** HASM_K, **c** HASM_I, **d** HASM_S

3.2 Comparison of the Sensitivity of HASM and HASM.MOD to the Selection of the Driving Field

Traditional HASM relies on other interpolation methods. To overcome this problem, HASM.MOD, which is built on the complete fundamental theorem of surface theory, introduces the equation satisfied by the mixed partial derivatives of the surface. In this section, the simulation performance of HASM and HASM.MOD with zero driving field is evaluated from a user perspective. First, the sensitivity of HASM and HASM.MOD to the selection of driving field is investigated for eight mathematical surfaces. Then, temperature and precipitation data are used to study the simulation accuracy of both methods with zero driving field.

3.2.1 Numerical Simulation

The simulation accuracies upon convergence with zero driving field of HASM and HASM.MOD are compared. The sampling ratio is 1/100. Table 3.3 shows the simulation root mean square error (RMSE).

With zero driving field, the accuracy of HASM.MOD is significantly higher than that of HASM for all eight surfaces. Figures 3.3 and 3.4 show the performance of HASM and HASM.MOD with zero driving field for the eight surfaces. The results demonstrate that except for f_2, f_5, and f_6, the results of traditional HASM are poor, particularly at the boundaries of the simulated area, where different degrees of oscillations occur for different surfaces. In contrast, the stimulation results of HASM.MOD for the eight surfaces are close to the real surfaces, and HASM.MOD shows a good simulation effect for all surfaces.

In summary, the performance of traditional HASM is significantly worse than that of HASM.MOD when zero driving field is used. For all eight surfaces, HASM.MOD with zero driving field demonstrates excellent simulation results, and the simulated surfaces are very close to the real surfaces. Thus, in practical applications, users no longer have to rely on other interpolation methods to generate the driving field; zero driving field is sufficient for HASM.MOD to produce good results.

Table 3.3 Simulation errors (RMSE) of HASM and HASM.MOD with zero driving field

Method	f_1	f_2	f_3	f_4	f_5	f_6	f_7	f_8
HASM.MOD	0.0003	0.0008	0.0003	0.0000	0.0000	0.1040	0.0000	0.0003
HASM	0.1456	0.0024	0.0843	0.0139	0.0002	0.1251	0.1052	0.1128

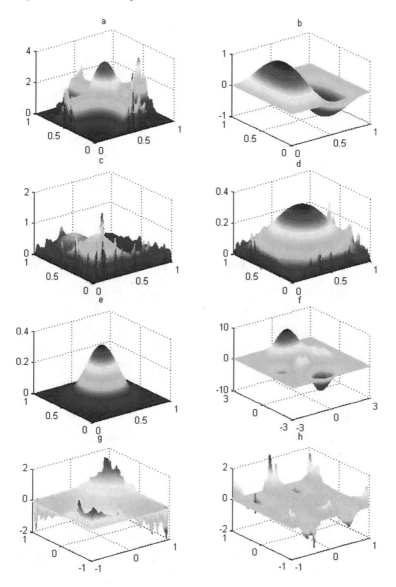

Fig. 3.3 Simulation results of traditional HASM for eight surfaces with zero driving field: **a** f1, **b** f2, **c** f3, **d** f4, I f5, **f** f6, **g** f7, **h** f8

3.2.2 Case Study

In this chapter, using temperature and precipitation as examples, the dependence of HASM and HASM.MOD on the selection of driving field is further evaluated. The simulation of temperature and precipitation is performed using the trend surface

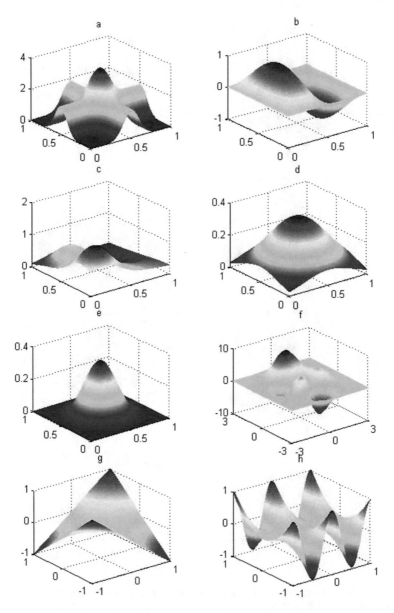

Fig. 3.4 Simulation results of HASM.MOD for eight surfaces with zero driving field: **a** f1, **b** f2, **c** f3, **d** f4, **e** f5, **f** f6, **g** f7, **h** f8

combined with HASM residual interpolation. To study the sensitivity of HASM and HASM.MOD to different driving fields, the multi-year average temperature in January and July from 1951 to 2010 is simulated. The differences between HASM and HASM.MOD with two different driving fields (kriging interpolation results and zero driving field) are presented in Table 3.4. The maximum (Max), minimum (Min), mean (Mean), and standard deviation (Std) of the differences are calculated using ArcGis 9.3. The last column in Table 3.4 shows the percentage of grids with a difference <1 °C with respect to all grids.

The simulated surfaces of HASM.MOD with different driving fields are similar, and the Std indicates that the surface obtained by taking the differences between the stimulated surfaces by the two HASM.MOD models (with two driving fields) becomes smooth. In contrast, the simulated surfaces of HASM with different driving fields differ significantly. The driving field has a significant influence on the performance of HASM, whereas HASM.MOD is almost independent of the driving field.

Figures 3.5 and 3.6 show the scatter plots of the measured values and values simulated by HASM and HASM.MOD with different driving fields in January and

Table 3.4 Differences between HASM and HASM.MOD regarding temperature simulation under different driving fields

Statistical indicators		Max	Min	Mean	Std	<1 °C (%)
January	HASM.MOD	1.1614	0.9476	0.009696	0.1697	99.98
	HASM	7.4078	10.5545	0.000831	0.6897	94.29
July	HASM.MOD	2.7861	1.5792	0.0110	0.2507	97.80
	HASM	0.0504	24.5604	0.0229	0.6801	94.09

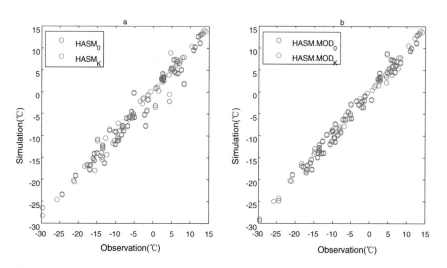

Fig. 3.5 Comparison of simulated and real temperature values in January: **a** HASM, **b** HASM.MOD

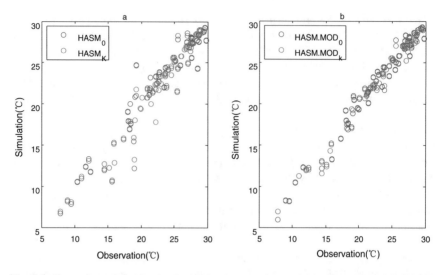

Fig. 3.6 Comparison of simulated and real temperature values in July: **a** HASM, **b** HASM.MOD

July, respectively, for 15% of the meteorological stations. $HASM_0$ and $HASM.MOD_0$ are the models with zero driving field, and $HASM_k$ and $HASM.MOD_k$ are the models with kriging interpolation results as the driving field. For both January and July, the simulation results of HASM.MOD with different driving fields exhibit small differences (Figs. 3.5b and 3.6b), while the simulation results of individual sites of HASM with different driving fields exhibit large differences (Figs. 3.5a and 3.6a).

Figures 3.7 and 3.8 show the distribution of simulated temperatures in January and July, respectively, under different driving fields before and after HASM improvement. HASM_K refers to the traditional HASM with kriging interpolation results as the driving field; HASM_0 is traditional HASM with zero driving field; HASM.MOD_K refers to HASM.MOD with kriging interpolation results as the driving field; HASM.MOD_0 is HASM.MOD with zero driving field. Figure 3.7 shows that when traditional HASM with different driving fields is used, there are noticeable differences in Xinjiang and Tibet regions. In comparison, the results of HASM.MOD with different driving fields are nearly the same.

For the average temperature in July of the traditional HASM, the differences with different driving fields are significant in Xinjiang, Tibet, and Inner Mongolia, and for HASM.MOD, the differences with different driving fields in Xinjiang are also significant; however, compared with that of traditional HASM, the sensitivity of HASM.MOD to the selection of driving field is much lower.

The simulation of precipitation is performed using geographically weighted regression (GWR) combined with HASM residual interpolation. Table 3.5 shows the differences between HASM and HASM.MOD with different driving fields.

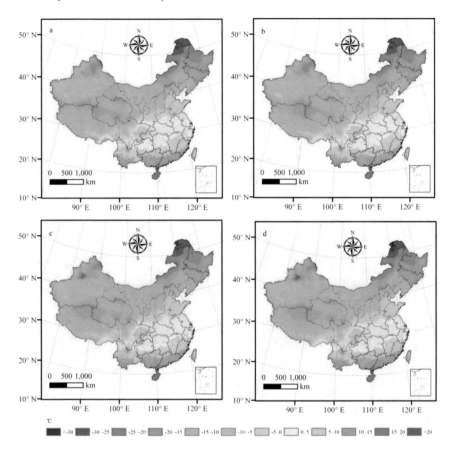

Fig. 3.7 Distribution of simulated temperatures in January: **a** HASM_K, **b** HASM_0, **c** HASM.MOD_K, **d** HASM.MOD_0

For the average precipitation in January and July of traditional HASM, the differences between different driving fields are large. In comparison, HASM.MOD is less sensitive to the selection of driving field.

Figures 3.9 and 3.10 show the differences between real precipitation and simulated precipitation of HASM and HASM.MOD with different driving fields at 15% of stations. The results of HASM with different driving fields are different at most of the stations, whereas the results of HASM.MOD with different driving fields are similar.

For comparison, Figs. 3.11 and 3.12 show the simulated precipitation distribution obtained by HASM and HASM.MOD with different driving fields. Although HASM and HASM.MOD with different driving fields are able to simulate the precipitation distribution, HASM with different driving fields shows distinct local differences, such as those in Xinjiang and Tibet in January (Fig. 3.11a, b). In contrast, the results of HASM.MOD with different driving fields are very close.

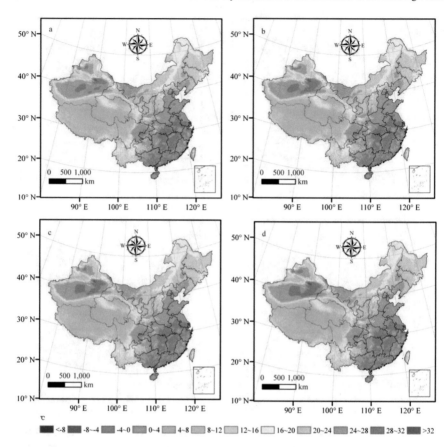

Fig. 3.8 Distribution of simulated temperatures in July: **a** HASM_K, **b** HASM_0, **c** HASM.MOD_K, **d** HASM.MOD_0

Table 3.5 Differences in simulated precipitation of HASM and HASM.MOD with different driving fields

Statistical indicators		Max	Min	Mean	Std	<1 mm (%)	<5 mm (%)
January	HASM.MOD	72.3428	9.5920	0.0314	0.9979	92.44	99.67
	HASM	147.1725	79.4179	0.0207	1.9125	89.05	99.17
July	HASM.MOD	35.2520	36.3376	0.2599	3.4402	44.82	88.77
	HASM	358.3737	61.6597	0.1452	6.0821	41.74	83.32

Based on the simulation results of temperature and precipitation, the results of HASM with different driving fields vary, while HASM.MOD is not affected by the driving field. Compared with the mathematical surfaces, the influencing factors of temperature and precipitation are complex and diverse, and HASM is used only to correct the residuals after removing the trend; therefore, the sensitivity of HASM to

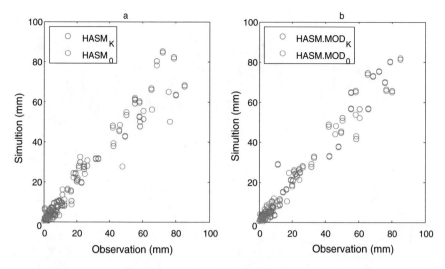

Fig. 3.9 Comparison of simulated and real precipitation in January: **a** HASM, **b** HASM.MOD

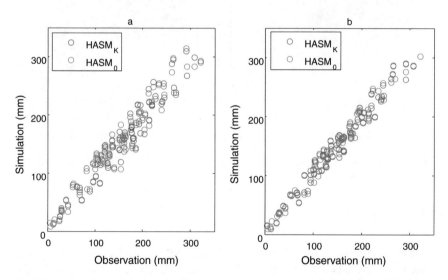

Fig. 3.10 Comparison of simulated and real precipitation in July: **a** HASM, **b** HASM.MOD

the driving field is not as significant as that in simulation of mathematical surfaces. Nonetheless, the influence of the driving field on the results of traditional HASM can still be observed from local details.

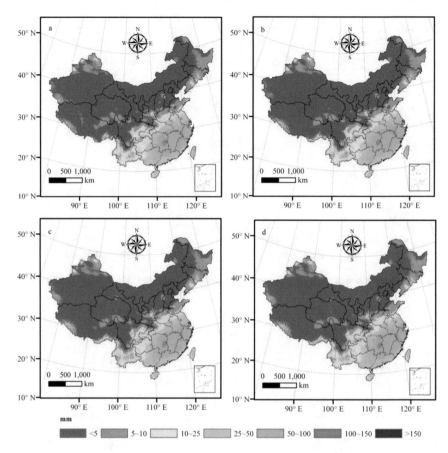

Fig. 3.11 National precipitation distribution in January: **a** HASM_K, **b** HASM_0, **c** HASM.MOD_K, **d** HASM.MOD_0

3.3 Summary

The main difference between HASM.MOD and traditional HASM is the introduction of the mixed partial derivative term in the Gaussian equations, resulting in different algebraic equations to be solved by HASM and HASM.MOD. Although the coefficient matrix w_{HASM} of the algebraic equation of traditional HASM is symmetric positive definite, the matrix is close to the singular state, with a minimum eigenvalue on the order of $O(10^{-7})$. Therefore, in the presence of an error perturbation in the computation process, singularity of the coefficient matrix easily occurs, so it is difficult to converge to the solution using the conjugate gradient (CG) method. Moreover, the maximum eigenvalue of the coefficient matrix is on the order of $O(10^4)$; thus, the condition number of w_{HASM} is as large as $O(10^{11})$ means that the rounding

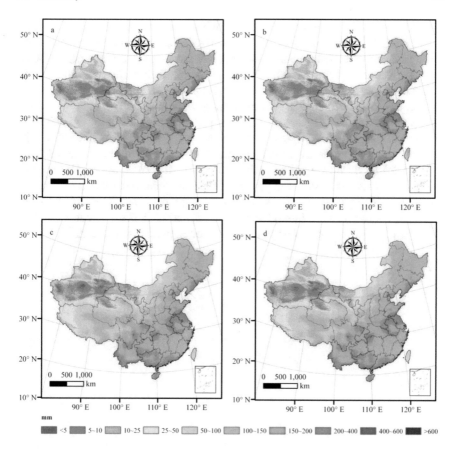

Fig. 3.12 National precipitation distribution in July: **a** HASM_K, **b** HASM_0, **c** HASM.MOD_K, **d** HASM.MOD_0

error or other errors , which can seriously affect the results. In contrast, the coefficient matrix $w_{HASM.MOD}$ of HASM.MOD has a minimum eigenvalue close to 1, and its condition number is approximately $O(10^6)$. Therefore, the solution can be obtained for any driving field using the CG method. Because of the introduction of mixed partial derivatives in HASM.MOD, the characteristics of the original surface are better restored than those of traditional HASM. Theoretical analysis shows that HASM.MOD is more stable than traditional HASM and that it is independent of the selection of driving field.

Since traditional HASM is dependent on a driving field generated by other methods, the application of the method is greatly restricted. In this chapter, through numerical simulation and case study, the sensitivity of traditional HASM and HASM.MOD to the selection of the driving field is evaluated. The following conclusions are drawn.

1. Traditional HASM is sensitive to the driving field to varying degrees. With zero driving field, the results of traditional HASM are not satisfactory. Since the results of other interpolation methods are used as the driving field, HASM is dependent on other methods and is merely a correction method for the other interpolation methods. In practical applications, repeated experiments are required for traditional HASM to obtain good simulation results.

2. HASM.MOD has a stable simulation performance due to its reliable theoretical foundation. The results show that the results of HASM.MOD with different driving fields are comparable. With zero driving field, the results of HASM.MOD are very close to the real values. Therefore, in practical applications, the user does not need provide the driving field; zero driving field can be used, which greatly simplifies the simulation process and makes HASM.MOD independent of other interpolation methods, so HASM.MOD can serve as a separate high-accuracy interpolation method in parallel to other methods.

Chapter 4
Influence of Sampling Information on the Performance of HASM.MOD

The modern high-accuracy surface modelling (HASM) method (HASM.MOD) is a spatiotemporal dynamic simulation method with global approximate data as the driving field and local high-precision data as the optimal control condition. It simulates the original surface based on the partial differential equations of the surface in differential geometry and the information of some points on the surface. If only solving the differential equations is considered, the results are infinite, and the corresponding linear algebraic equation has infinite solutions. However, the sampling information helps to determine a finite solution. Therefore, the sampling point information is bound to have a significant influence on the simulation accuracy of HASM.MOD. In this chapter, the effect of sampling information on the performance of HASM.MOD is investigated from the aspects of sampling ratio and sampling error.

4.1 Influence of the Sampling Ratio on Simulation Accuracy of HASM.MOD

Spatial sampling is an important way to estimate the statistical parameters of regional attribute characteristics and to build spatial variation analysis models. The number of sampling points is the key factor in determining the sampling cost and accuracy (De Gruijter et al. 2006). For a given computational grid size, i.e., a fixed spatial resolution, the sampling ratio is defined as the ratio of the number of sampling points to the number of computational grids. In traditional HASM, the influence of the sampling ratio on the simulation accuracy is negligible (Yue and Du 2005). In this chapter, the influence of the sampling ratio on the performance of HASM.MOD is investigated.

4.1.1 Numerical Simulation

The changes in simulation accuracy of HASM.MOD with increasing sampling density with the Gaussian surface as an example are shown in Fig. 4.1. The sampling ratio is set to 1, 2, ..., 5, ..., 10, 20, 30, 40 and 50%.

The simulation error of HASM.MOD decreases rapidly as the sampling density increases. When the sampling ratio is 10%, the simulation results are already accurate. As the sampling ratio continues to increase, the simulation accuracy of HASM.MOD improves slightly. Figure 4.2 shows the simulation results of the Gaussian surface when the sampling ratio is 1 and 50%. Figure 4.3 shows the error distribution between the simulated and real Gaussian surface. The results show that the simulation error of HASM.MOD is large when the sampling ratio is 1% and that the error distribution ranges from −0.13 to 0.12.

Figure 4.4 shows the simulation error of HASM.MOD for the saddle surface at different sampling ratios. With increasing sampling ratio, the simulation error decreases significantly, and when the sampling ratio reaches 10%, the influence of sampling ratio on the simulation error decreases.

Figure 4.5 shows the saddle surface simulated by HASM.MOD at sampling ratios of 1 and 50%. Figure 4.6 shows the difference between the simulated and real saddle surfaces. The results indicate that the simulation error of HASM.MOD is large when the sampling ratio is low and that as the sampling ratio increases, the simulated value is almost equal to the real value.

For mathematical surface f_1 : $f_1(x, y) = e^{\left\{-\frac{(5-10x)^2}{2}\right\}} + 0.75e^{\left\{-\frac{(5-10y)^2}{2}\right\}} + 0.75e^{\left\{-\frac{(5-10x)^2}{2}\right\}} + e^{\left\{-\frac{(5-10y)^2}{2}\right\}}$, Fig. 4.7 shows the changes in

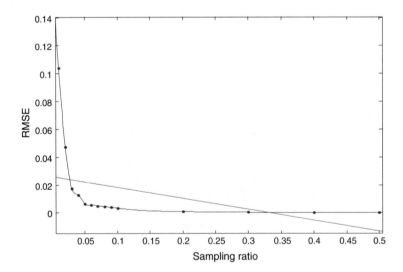

Fig. 4.1 Influence of the sampling ratio on the simulation results of the Gaussian surface

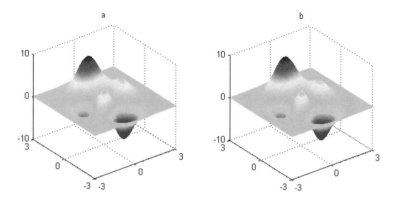

Fig. 4.2 Simulation of the Gaussian surface with different sampling ratios. **a** Sampling ratio of 1%. **b** Sampling ratio of 50%

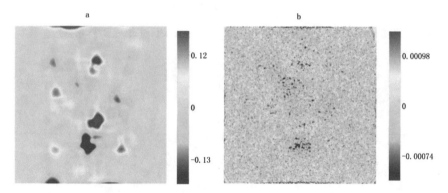

Fig. 4.3 Error distribution between simulated and real Gaussian surface. **a** Sampling ratio of 1%. **b** Sampling ratio of 50%

the simulation error of HASM.MOD with the sampling ratio. The simulation accuracy gradually increases as the sampling ratio increases. In addition, when the sampling ratio reaches 10%, the simulation accuracy of HASM.MOD is almost no longer influenced by the sampling ratio.

Figure 4.8 shows the performance of HASM.MOD at different sampling ratios. Figure 4.9 shows the error distribution between the simulated and real surface. At a low sampling ratio, the simulation error is large. When the sampling ratio is increased to 50%, the error between the stimulated surface of HASM.MOD and the real surface is in the range of −0.00001–0.00001.

The numerical experiment shows that as the number of sampling points increases, the HASM.MOD simulation accuracy increases. However, the influence of the sampling ratio on the performance of HASM.MOD is limited. When the sampling ratio reaches a certain threshold, the simulation accuracy of HASM.MOD improves

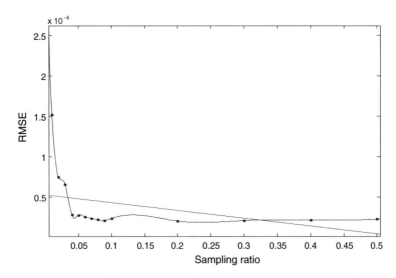

Fig. 4.4 Influence of the sampling ratio on the simulation results of the saddle surface

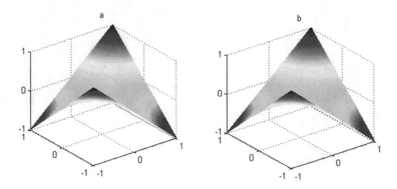

Fig. 4.5 Simulation of the saddle surface at different sampling ratios. **a** Sampling ratio of 1%.
b Sampling ratio of 50%

only slightly with increasing sampling ratio; thus, the influence of the sampling ratio
on the simulation accuracy of HASM.MOD decreases.

4.1.2 Case Study

In this section, taking temperature and precipitation as examples, the influence of
the sampling ratio on the simulation accuracy is analysed. First, the temperature
and precipitation data of 5% of 752 meteorological stations in China are selected as
the verification dataset, and then 30, 40, 50, ... and 90% of the remaining stations

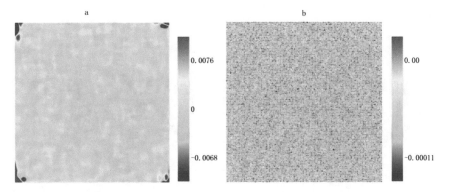

Fig. 4.6 Error distribution between the simulated and real values of the saddle surface. **a** Sampling ratio of 1%. **b** Sampling ratio of 50%

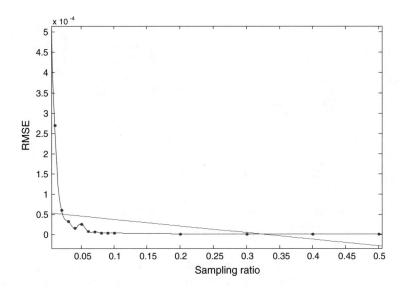

Fig. 4.7 Influence of the sampling ratio on the simulation results of surface f_1

are randomly selected as the sampling points for simulation. The experiment is repeated 10 times. Figure 4.10 shows the changes in the simulation error in the average temperature in January from 1951 to 2010 with the sampling ratio. The sampling ratio has a strong influence on the simulation results. When the sampling ratio is 60%, the results of HASM.MOD are less accurate than those of the 50% sampling ratio, which may be due to the randomness of sampling point selection and the limited number of tests. Overall, the simulation accuracy of HASM.MOD tends to increase as the sampling ratio increases.

Figure 4.11 shows the average temperature in January simulated by HASM.MOD when the sampling ratio is 30 and 90%. The results show that the performance of

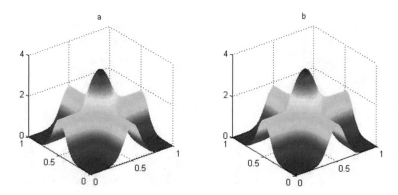

Fig. 4.8 Simulation of surface f1 at different sampling ratios. **a** Sampling ratio of 1%. **b** Sampling ratio of 50%

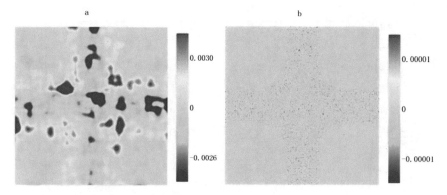

Fig. 4.9 Error distribution between the simulated and real values of surface f_1. **a** Sampling ratio of 1%. **b** Sampling ratio of 50%

HASM.MOD is different at different sampling ratios. Especially in Xinjiang and Inner Mongolia, HASM.MOD performs well at a 90% sampling ratio, and the simulation values equal the real values. Figure 4.12 shows the differences between the simulated and measured values for the 5% verification stations with different sampling ratios; the simulation error of HASM.MOD is small when the number of sampling points is large.

Figure 4.13 shows the changes in the simulation error of HASM.MOD with sampling ratio for the average temperature in July. As the sampling ratio increases, the simulation accuracy of HASM.MOD tends to increase. Figure 4.14 shows the simulation results of HASM.MOD for the average temperature in July when the sampling ratio is 30 and 90%. The results show that the border regions or regions with sparse stations, such as Xinjiang and Inner Mongolia, are still the regions that exhibit obvious differences between the two sampling ratios. Figure 4.15 shows the differences between the simulated and measured values in the 5% verification stations.

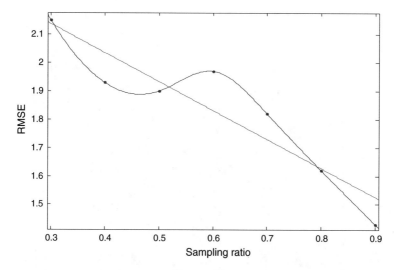

Fig. 4.10 Influence of the sampling ratio on the simulation results of average temperature in January

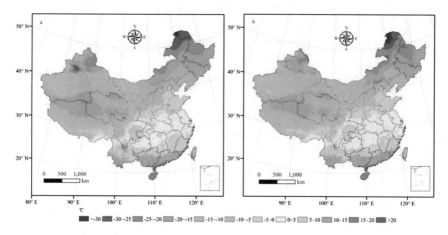

Fig. 4.11 HASM.MOD simulation of January temperature at different sampling ratios. **a** Sampling ratio of 30%. **b** Sampling ratio of 90%

Since the finite-difference discretization of HASM.MOD uses the points near the grid points, at some stations, the performance of HASM.MOD at a sampling ratio of 90% is less accurate than that of a sampling ratio of 30%. However, generally speaking, the higher the sampling ratio is, the higher the simulation accuracy of HASM.MOD.

In terms of precipitation, the influence of the sampling ratio on the performance of HASM.MOD is investigated. According to the simulation results of the average precipitation in January, the simulation accuracy appears to increase or decrease to

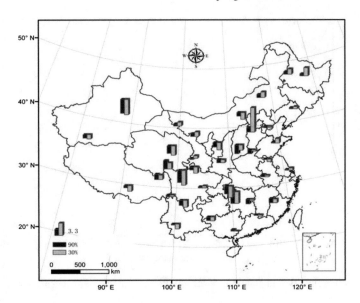

Fig. 4.12 Differences between simulated and real temperatures in January at different sampling ratios

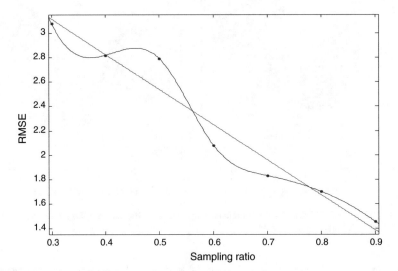

Fig. 4.13 Influence of the sampling ratio on the simulation results of the average temperature in July

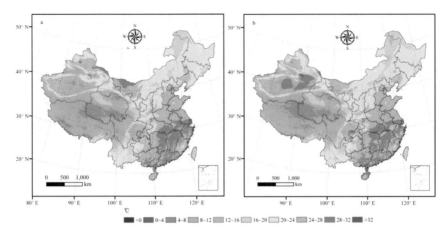

Fig. 4.14 Simulation of July temperature by HASM.MOD at different sampling ratios. **a** A sampling ratio of 30%. **b** A sampling ratio of 90%

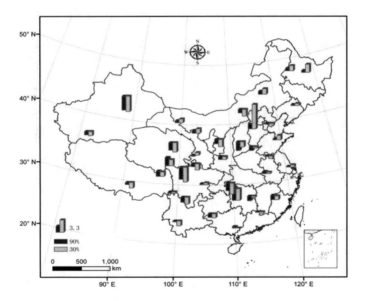

Fig. 4.15 Difference between simulated and measured values of July temperatures at different sampling ratios

varying degrees with increasing sampling ratio. This is mainly due to the randomness of sampling points and the fact that the simulation accuracy of flat terrain is higher than that of steep terrain. On the whole, the simulation accuracy of HASM.MOD tends to increase gradually as the sampling ratio increases (Fig. 4.16).

Figure 4.17 shows the simulation results of average precipitation in January at different sampling ratios. In northern Xinjiang, southwest Tibet, Baqing county,

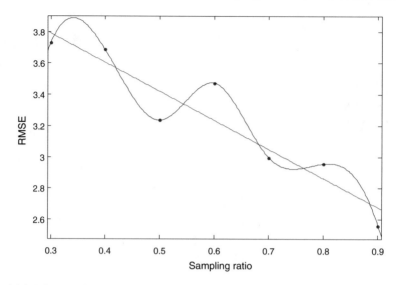

Fig. 4.16 Influence of the sampling ratio on the simulation results of precipitation in January

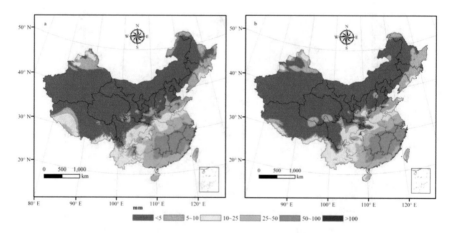

Fig. 4.17 Simulation of precipitation in January by HASM.MOD with different sampling ratios.
a A sampling ratio of 30%. **b** A sampling ratio of 90%

Shiqu county in Tibet, southern Yunnan, southern Jilin, the junction of Shaanxi
and Hubei, and central Shaanxi, the differences between the two different sampling
ratios are relatively large. Compared with the results with a low sampling ratio, the
simulation results are closer to the real values when the sampling ratio is 90%.

Figure 4.18 shows the differences between the simulated and measured values
of the average precipitation in January. At some stations, the simulated values by
HASM.MOD with a sampling ratio of 30% are higher than the measured values,
and the simulated values by HASM.MOD with a sampling ratio of 90% are lower

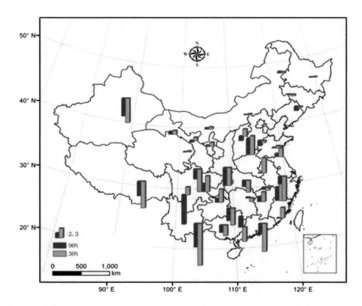

Fig. 4.18 Differences between simulated and measured values of precipitation in January at different sampling ratios

than the measured values. HASM.MOD simulation at 90% sampling ratio is more accurate than that of 30% sampling ratio.

Similarly, for July, which typically has more precipitation, the influence of sampling ratio on the performance of HASM.MOD is investigated. Figure 4.19 shows

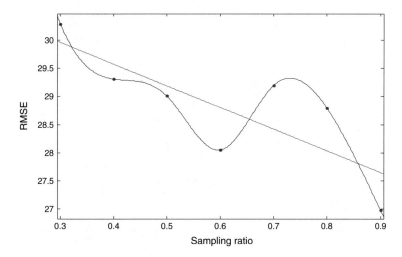

Fig. 4.19 Influence of the sampling ratio on the simulation results of precipitation in July

that, with a limited number of tests, the influence of the sampling ratio on the simulation accuracy of HASM.MOD is not obvious. However, generally speaking, with increasing sampling ratio, the overall simulation accuracy of HASM.MOD tends to improve.

Figures 4.20 and 4.21 show the performance of HASM.MOD when the sampling

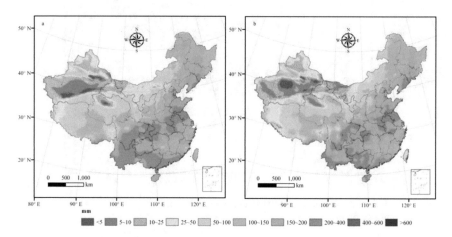

Fig. 4.20 Simulation of precipitation in July by HASM.MOD at different sampling ratios. **a** A sampling ratio of 30%. **b** A sampling ratio of 90%

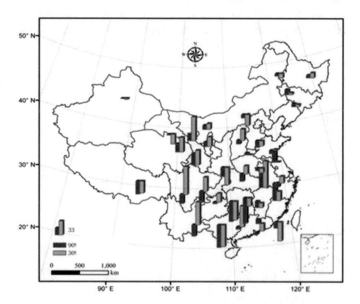

Fig. 4.21 Differences between the simulated and measured values of precipitation in July at different sampling ratios

ratio is 30 and 90%, respectively. At a sampling ratio of 90%, the results of HASM.MOD demonstrate more detailed features. In the Hengduan Mountains and the surrounding areas, the simulation results in Fig. 4.20b are closer to the real values. The differences between the two sampling ratios are large in the border regions and regions with sparse stations. Figure 4.21 shows that the simulated values of HASM.MOD are closer to the measured values when there are more sampling points.

In summary, the sampling ratio has an important influence on the simulation accuracy of HASM.MOD. With increasing sampling ratio, the simulation accuracy increases. Moreover, the representativeness of sampling points has certain influence on the simulation accuracy as well. Without a significant increase in the sampling ratio, the simulation accuracy of HASM.MOD may not be necessarily improved. Therefore, the location information of sampling points should be taken into account when increasing the sampling points.

4.2 Influence of the Sampling Error on the Simulation Accuracy of HASM.MOD

The quality of sampling data has an important influence on the performance of HASM.MOD. In this section, the influence of the sampling error on the simulation accuracy of HASM.MOD is studied.

4.2.1 Numerical Simulation

With a sampling ratio of 10%, random errors with a standard deviation of 5.77×10^{-6}, 2.88×10^{-5}, 5.77×10^{-5}, 2.88×10^{-4}, 5.77×10^{-4}, 2.88×10^{-3}, 5.77×10^{-3}, 2.88×10^{-2}, and 5.77×10^{-2} are separately added to the sampling points. The changes in the simulation error of HASM.MOD with the sampling error for the Gaussian surface as an example are shown in Fig. 4.22. For HASM.MOD, the simulation error increases almost linearly with increasing sampling error.

Figure 4.23 shows the Gaussian surface simulated by HASM.MOD when the sampling errors are 5.77×10^{-6} and 5.77×10^{-2}. When the sampling error is large, the simulated surface of HASM.MOD oscillates, and the accuracy is obviously lower than that when the sampling error is small.

Figure 4.24 shows the differences between the simulated and real Gaussian surfaces. When the sampling error is 5.77×10^{-6}, the simulation error is between -0.0044 and 0.0032. When the sampling error is large, the simulation results of HASM.MOD are quite different from the real results, with the difference ranging from -0.45 to 1.28.

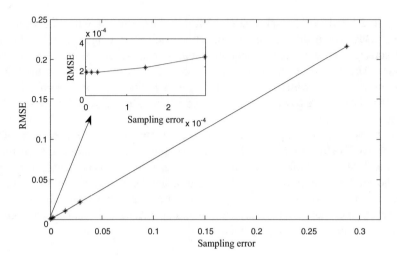

Fig. 4.22 Influence of the sampling error on the simulation results of the Gaussian surface

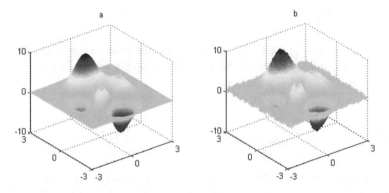

Fig. 4.23 Simulation of the Gaussian surface at different sampling errors. **a** A sampling error of 0.0000577. **b** A sampling error of 0.0577

For the saddle surface, the simulation results of HASM.MOD with different sampling errors are investigated. Figure 4.25 shows the changes in the HASM.MOD simulation error with the sampling error. With increasing sampling error, the simulation error of HASM.MOD increases gradually.

Figure 4.26 shows the simulation of HASM.MOD for the saddle surface when the errors of 5.77×10^{-6} and 5.77×10^{-2} are added to the sampling points. Figure 4.27 shows the error distribution between the simulated and real saddle surfaces. When the sampling error is large, the simulation results are not accurate, and the simulated values oscillate, with differences ranging from -0.63 to 1.29.

Figure 4.28 illustrates the changes in the simulation error with the sampling error. Figure 4.29 shows the simulated surface f_1 when the sampling errors are 5.77×10^{-6}

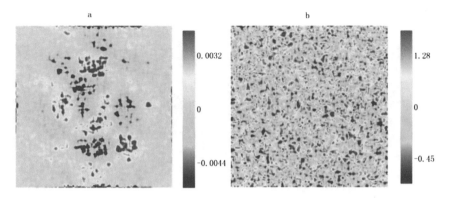

Fig. 4.24 Differences between the simulated and real Gaussian surfaces at different sampling errors. **a** A sampling error of 0.0000577. **b** A sampling error of 0.0577

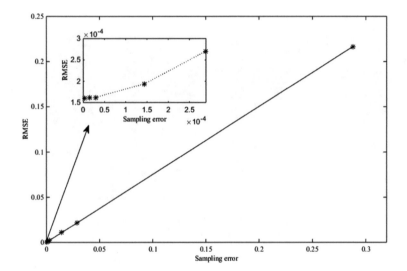

Fig. 4.25 Influence of the sampling error on the simulation accuracy of the saddle surface

and 5.77×10^{-2}. Figure 4.30 shows the differences between the simulated and real surfaces. With increasing sampling error, the simulation error of HASM.MOD gradually increases. When the sampling error increases to 5.77×10^{-2}, the difference between the simulated and real values ranges from -0.57 to 1.56.

The simulation results of the above three mathematical surfaces show that with increasing sampling error, the simulation error of HASM.MOD increases almost linearly. As seen from the enlarged insets, small errors at the sampling points have little effect on the final results, yet as the sampling error increases, the error in the simulation results increases significantly.

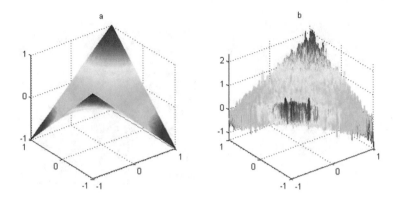

Fig. 4.26 Simulation of the saddle surface at different sampling errors. **a** A sampling error of 0.00000577. **b** A sampling error of 0.0577

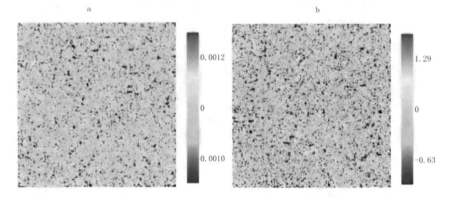

Fig. 4.27 Differences between the simulated and real saddle surfaces at different sampling errors. **a** A sampling error of 0.00000577. **b** A sampling error of 0.0577

4.2.2 Case Study

From the 752 meteorological stations in China, the meteorological data of 90% of the stations are randomly selected as simulation data, and the other 10% are used as test data. Errors with the standard deviations of 2.88×10^{-6}, 1.44×10^{-5}, 2.88×10^{-5}, 1.44×10^{-4}, 2.88×10^{-4}, 1.44×10^{-3}, 2.88×10^{-3}, 1.44×10^{-2}, 2.88×10^{-2}, 1.44×10^{-1}, 2.88×10^{-1}, 7.02×10^{-1}, 1.41 and 2.87 are added to the simulation dataset. The influence of sampling error on the simulation results of HASM.MOD for the average temperature in January as an example is shown in Fig. 4.31. There is a small increase in simulation error when the sampling error is small. As the sampling error increases, the simulation error increases significantly.

Figure 4.32 shows the simulation results of HASM.MOD for the average temperature in January when the sampling errors are 2.88×10^{-6} and 2.87. With a large

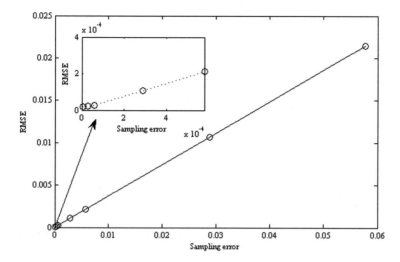

Fig. 4.28 Influence of the sampling error on the simulation results of surface f_1

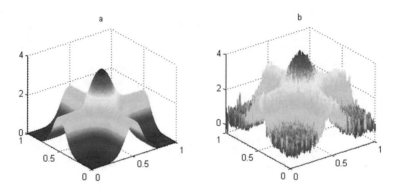

Fig. 4.29 Simulation of surface f_1 at different sampling errors. **a** A sampling error of 0.00000577. **b** A sampling error of 0.0577

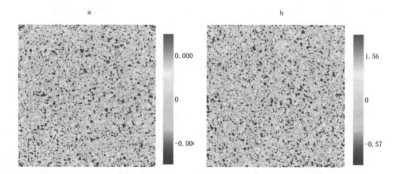

Fig. 4.30 Differences between the simulated and the real values of surface f_1 with different sampling errors. **a** A sampling error of 0.00000577. **b** A sampling error of 0.0577

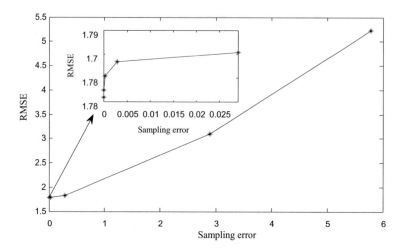

Fig. 4.31 Influence of the sampling error on the simulation results of average temperature in January

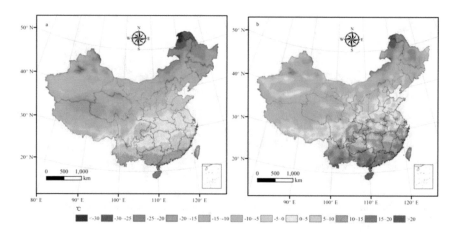

Fig. 4.32 Simulation of the average temperature in January with different sampling errors. **a** A sampling error of 0.00000288. **b** A sampling error of 2.87

sampling error, the simulated values of HASM.MOD oscillate significantly, and the simulated values differ significantly from the real values.

Figure 4.33 shows the differences between the simulated and measured values of temperature at the 10% validation stations with different sampling errors. When the sampling error is large, the simulation error is significantly large, while the simulation error is relatively small when sampling error is 2.88×10^{-6}.

According to Fig. 4.34, for the average temperature in July, a small sampling error causes a slight decrease in the simulation error of HASM.MOD, and as the sampling error continuously increases, the simulation error increases significantly. Therefore,

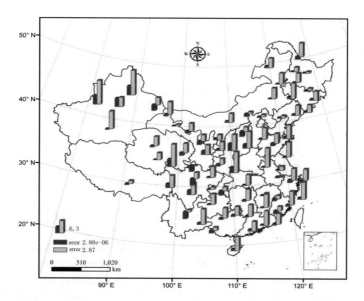

Fig. 4.33 Differences between the simulated and measured temperature in January with different sampling errors

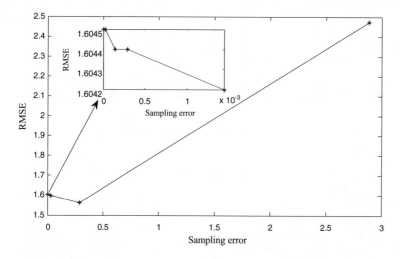

Fig. 4.34 Influence of the sampling error on the average temperature in July

the sampling error does not always increase the simulation error of HASM.MOD due to the original error of the sampling points being under the influence of various factors. Thus, adding extra error on the top of the original error does not necessarily make the value of sampling point differ more from the real value. As the sampling error increases, the simulation accuracy is significantly affected.

Figure 4.35 shows the simulation of the temperature in July when the sampling errors are 2.88×10^{-6} (Fig. 4.35a) and 2.87 (Fig. 4.35b). Comparing Fig. 4.35a with Fig. 4.35b shows that the simulated values with a low sampling error are closer to the real values. When the sampling error is 2.87, the simulation values oscillate, and the simulation results are quite different from the real values.

Figure 4.36 shows the differences between the simulated and measured values for the 10% validation stations at different sampling errors. When the sampling error is large, the simulated values are quite different from the measured values.

For the average precipitation in January, Fig. 4.37 shows the influence of the sampling error on the simulation results of HASM.MOD on a national scale. The simulation accuracy of HASM.MOD gradually decreases with increasing sampling error.

Figure 4.38 shows the simulation results of HASM.MOD for precipitation in January when the sampling error is 2.88×10^{-6} (Fig. 4.38a) and 2.87 (Fig. 4.38b). When the sampling error is small, the simulation results of HASM.MOD in Heilongjiang are different from the real values. When the sampling error is large, the simulation results of HASM.MOD are very different from the real values. Figure 4.39 shows the difference between the simulated and measured values for the 10% validation stations. When the sampling error is large, the simulated values are significantly different from the measured values.

Figure 4.40 shows the changes in the simulation error of the average precipitation in July with the sampling error. The results show that with increasing sampling error, the simulation error of HASM.MOD increases gradually. Figure 4.41 shows the simulation results of HASM.MOD for July precipitation when the sampling error is 2.88×10^{-6} (Fig. 4.40a) and 2.87 (Fig. 4.40b). The simulation results with the two sampling errors are significantly different, especially in the Xinjiang area.

Figure 4.42 shows the differences between the simulated and measured values for

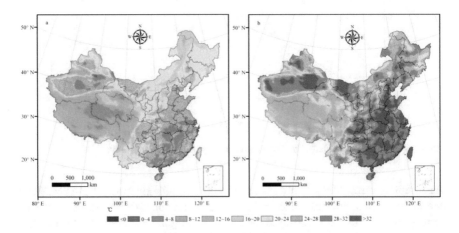

Fig. 4.35 Simulation results of temperature in July with different sampling errors. **a** A sampling error of 0.0000288. **b** A sampling error of 2.87

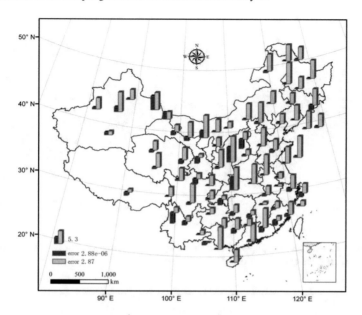

Fig. 4.36 Differences between the simulated and measured values of temperature in July at different sampling errors

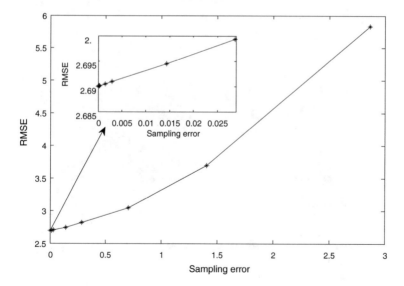

Fig. 4.37 Influence of the sampling error on the simulation results of precipitation in January

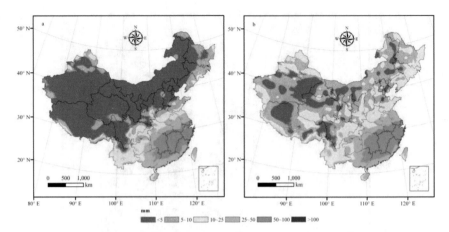

Fig. 4.38 Simulation of precipitation in January with different sampling errors. **a** A sampling error of 0.0000288. **b** A sampling error of 2.87

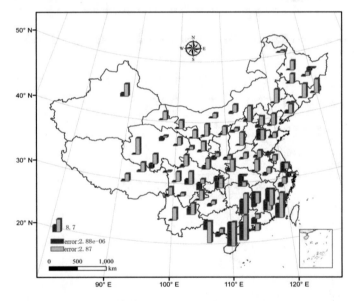

Fig. 4.39 Differences between the simulated and measured values of precipitation in January at different sampling errors

the 10% validation stations. The simulation error of HASM.MOD is small when the sampling error is small.

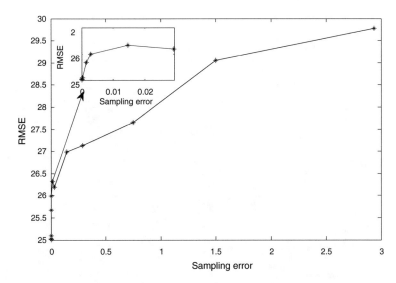

Fig. 4.40 Influence of the sampling error on the simulation results of precipitation in July

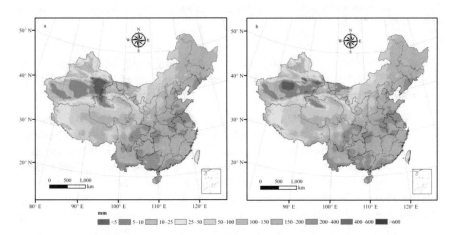

Fig. 4.41 Simulation of the precipitation in January with different sampling errors. **a** A sampling error of 0.00000288. **b** A sampling error of 2.87

4.3 Summary

Since HASM.MOD is a high precision surface modelling method with global approximation data as the driving field and local high precision data as the optimal control condition, the sampling information is crucial to the simulation results.

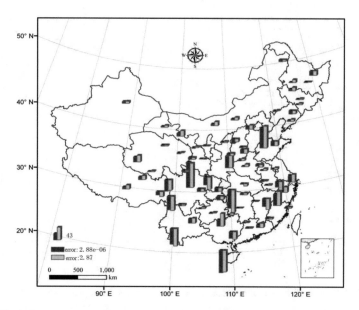

Fig. 4.42 Differences between the simulated and measured values of precipitation in July at different sampling errors

The influence of the sampling ratio and sampling error on the accuracy of HASM.MOD is investigated through numerical simulation and a case study. The conclusions are as follows:

(1) According to the results of numerical simulations, with increasing sampling ratio, the simulation accuracy of HASM.MOD first increases and then becomes stable. Due to various factors in practical applications, a certain threshold can be set for the sampling ratio when it is difficult to obtain a sufficient number of sample points. The simulation of temperature and precipitation shows that trend of the simulation accuracy is different from that of the numerical experiment and that the simulation accuracy increases continuously with increasing sampling ratio. Because the currently available sampling points (752 stations) are relatively insufficient for a national scale simulation, it is hard to capture the effect of the number of sampling points on the simulation results given limited sampling information. If there are enough sampling points, the influence of the sampling ratio on the HASM.MOD simulation results may be similar to that of mathematical surfaces. However, based on the current 752 meteorological stations in China, the case study indicates that the accuracy of the HASM.MOD simulation gradually increases with an increasing number of sampling points.

(2) Compared with the sampling ratio, the sampling error has a stronger influence on the simulation results of HASM.MOD. The influence of the sampling error is significantly high when the real values of the surface vary over a large range. When the range of the real values of the surface is small, the influence of the

sampling error is small. The case study shows that small errors in the sampling values do not necessarily increase the simulation error of HASM.MOD and that the simulation error changes only in a small range. As the sampling error continues to increase, the simulation error increases significantly.

References

De Gruijter JJ, Brus DJ, Bierkens MFP, et al. 2006.Sampling for natural resource monitoring. Berlin: Springer.
Yue TX, Du ZP. 2005. High precision surface modeling: the core module of new generation GIS and CAD.Progress in Natural Science, 15(3): 73–82.

Chapter 5
Fast Computation and Parallel Computing of HASM.MOD

Based on the principal theorem of surface theory, the modern high-accuracy surface modelling (HASM) method (HASM.MOD) transforms surface simulation into solving large-scale sparse algebraic equations after finite-difference discretization of the partial differential equations of the surface. Experience has shown that the time consumption in solving linear algebraic equations often accounts for a large proportion of the total computation time; hence, it is particularly important for HASM.MOD to explore efficient solving methods. Through the analysis of the structural characteristics of the coefficient matrix of the HASM.MOD equations, the optimal preconditioning operator based on the conjugate gradient (CG) method is selected to transform the original equations into linear equations with good properties, thereby increasing the convergence speed of HASM.MOD and shortening the processing time. Moreover, in the message passing interface (MPI) parallel programming environment, the parallel computation of HASM.MOD is realized.

5.1 Fast Computation Method for HASM.MOD

Large-scale linear algebraic equations are generally solved by iteration methods, and the convergence rate determines the success of the methods. Classical iteration methods include Jacobi iteration, Gauss–Seidel iteration, and successive over-relaxation (SOR) iteration. At each iteration step, Jacobi iteration uses all the components of the results obtained at the previous step, which not only occupies an extensive amount of memory but also converges rather slowly. On the basis of Jacobi iteration, Gauss–Seidel iteration fully utilizes the newly calculated components, and the computation speed and accuracy are improved. However, the convergence of certain problems is slow. The SOR iteration, as an extension of the Gauss–Seidel iteration, is an effective approach for solving large sparse equations. A relaxation parameter is introduced to increase the computation speed. When the relaxation parameter is 1, the SOR iteration is Gauss–Seidel iteration. However, in practice, the selection

N. Zhao and T. Yue, *High Accuracy Surface Modeling Method: The Robustness*,
https://doi.org/10.1007/978-981-16-4027-8_5

of the relaxation parameter is often a challenging problem. The CG method is an effective method for solving large-scale sparse symmetric positive definite systems (Golub and Van loan 2009). Based on the steepest descent method, the CG method seeks the direction with the fastest decrease rate in the error.

The algebraic equation of HASM.MOD is as follows:

$$\bar{\mathbf{W}}x = \bar{\mathbf{v}} \tag{5.1}$$

The solving process of the CG method is as follows:

Algorithm 1: Conjugate Gradient (CG) (Golub and Loan 2009)

$$\text{Given an initial } x_0,$$
$$k = 0, r_0 = \bar{\mathbf{v}} - \bar{\mathbf{W}}x_0,$$
$$\text{while } (r_k \neq 0)$$
$$k=k+1$$
$$\text{If } k = 1$$
$$p_1 = r_0$$
$$\text{else}$$
$$\beta_k = r_{k-1}^T r_{k-1} / r_{k-2}^T r_{k-2}$$
$$p_k = r_{k-1} + \beta_k p_{k-1}$$
$$\text{end}$$
$$\alpha_k = r_{k-1}^T r_{k-1} / p_k^T \bar{\mathbf{W}} p_k$$
$$x_k = x_{k-1} + \alpha_k p_k$$
$$r_k = r_{k-1} - \alpha_k \bar{\mathbf{W}} p_k$$
$$\text{end}$$
$$x = x_{k-1}$$

During the implementation of the CG method, it is necessary to calculate the product of the matrix and vector $w = \bar{\mathbf{W}} p_k$, the inner product of the two vectors $r_{k-1}^T r_{k-1}$ and $p_k^T w$, and the product of the three numbers and vectors $\beta_k p_{k-1}$, $\alpha_k p_k$ and $\alpha_k w$. Therefore, the time complexity is $O(n^2)$. Four one-dimensional vectors $w = \bar{\mathbf{W}} p_k$, x_k, r_k, and p_k need to be stored, and the space complexity is $O(n)$.

The CG method has the following properties:

(1) $p_i^T r_j = 0$, $(0 \leq i < j \leq k$, k denotes the current iteration number, p_i is the search direction of the CG method, and r_j represents the residual vector of the iteration).
(2) $r_i^T r_j = 0$, $(i \neq j, i, j \leq k)$.
(3) $p_i^T \bar{\mathbf{W}} p_j = 0$, $(i \neq j)$

Theoretically, the CG method can yield the exact solution of n-order linear equations only through n iterations. However, in practice, due to computer round-off errors and the ill-conditioned properties of coefficient matrix $\bar{\mathbf{W}}$, the orthogonality of $p_i, i = 1, 2, \cdots$ and $r_i, i = 1, 2, \cdots$ disappears as the number of iterations increases during the computation process of the CG method. If $\bar{\mathbf{W}} \in \mathbf{R}^{n \times n}$, then the CG method exhibits the following error when solving Eq. (5.1):

$$\left\| x^{(k)} - x^* \right\|_V \leq 2 \left(\frac{\sqrt{K} - 1}{\sqrt{K} + 1} \right)^k \left\| x^{(0)} - x^* \right\|_V \qquad (5.2)$$

where $x^{(k)}$ is the result of the k-th iteration; x^* denotes the exact solution; $x^{(0)}$ represents the initial value; K is the conditional number of the matrix $\bar{\mathbf{W}}$ (the quantity describing the ill-conditioned properties of the matrix), in which $K = \frac{\lambda_n}{\lambda_1} = \left\| \bar{\mathbf{W}}^{-1} \right\|_v \left\| \bar{\mathbf{W}} \right\|_v$, and $V = 1, 2,$ or ∞. λ_n and λ_1 are the maximum and minimum eigenvalues of matrix $\bar{\mathbf{W}}$, respectively. Consequently, if the conditional number K of $\bar{\mathbf{W}}$ is very large or the difference between λ_n and λ_1 is too large, the convergence of the method could be rather slow. In other words, the convergence rate of the CG method is determined by the conditional number of $\bar{\mathbf{W}}$, or more generally, by the distribution of eigenvalues of $\bar{\mathbf{W}}$. Figure 5.1 shows the convergence diagram of the CG method when K is small (a) and large (b).

According to the equation of the conditional number, when $K \geq 1$ and matrix $\bar{\mathbf{W}}$ is a unit matrix $\mathbf{I}$, $K = 1$. If $\bar{\mathbf{W}} = \mathbf{I} + \mathbf{B}$ and the order of $\mathbf{B}$ is r, then $x^{(r+1)} = x^*$ is obtained by the CG method. Thus, when matrix $\bar{\mathbf{W}}$ is close to a unit matrix $\mathbf{I}$, the CG method converges quickly. However, the conditional number K of the coefficient matrix $\bar{\mathbf{W}}$ in HASM.MOD Eq. (5.1) is very large and can reach the order of $O(10^6)$.

To improve the matrix $\bar{\mathbf{W}}$ and accelerate convergence, Meijerink and van der Vorst proposed the preconditioned CG (PCG) method in 1977. The PCG method attempts to find a preconditioned matrix $\mathbf{M}$ to minimize the conditional number of $\mathbf{M} * \bar{\mathbf{W}}$

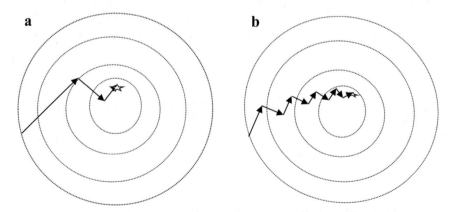

Fig. 5.1 Convergence diagram of the CG method. **a** A small K value. **b** A large K value

and transform $\bar{\mathbf{W}}x = \bar{\mathbf{v}}$ into a set of equations $\mathbf{M}\bar{\mathbf{W}}x = \mathbf{M}\bar{\mathbf{W}}\mathbf{N}$ that has the same solution. The steps of the PCG method are as follows:

Algorithm 2: Preconditioned Conjugate Gradient (PCG) (Golub and Loan 2009)

$$\text{Given an initial } x_0,$$
$$k = 0, r_0 = \bar{\mathbf{v}} - \bar{\mathbf{W}}x_0,$$
$$\text{while } (r_k \neq 0)$$
$$\text{solve } \mathbf{M}z_k = \mathbf{r}_k$$
$$k=k+1$$
$$\text{If } k = 1$$
$$p_1 = z_0$$
$$\text{else}$$
$$\beta_k = r_{k-1}^T z_{k-1} / r_{k-2}^T z_{k-2}$$
$$p_k = z_{k-1} + \beta_k p_{k-1}$$
$$\text{end}$$
$$\alpha_k = r_{k-1}^T z_{k-1} / p_k^T \bar{\mathbf{W}} p_k$$
$$x_k = x_{k-1} + \alpha_k p_k$$
$$r_k = r_{k-1} - \alpha_k \bar{\mathbf{W}} p_k$$
$$\text{end}$$
$$x = x_{k-1}$$

In the PCG method, $\mathbf{M}z_k = r_k$ is solved during each iteration. Therefore, based on differences between the preconditioned matrix M, the time complexity is $O(n^3)$ or $O(n^2)$, and the space complexity is $O(n^2)$ or $O(n)$.

In this chapter, according to the characteristics of the coefficient matrix in the HASM.MOD equations, through the selection of different preconditioned matrices **M**, the transformed algebraic system achieves a high convergence speed. The PCG method is essentially the application of the CG method to the transformed equations:

$$\bar{\mathbf{A}}x = \bar{\mathbf{b}}, \tag{5.3}$$

where $\bar{\mathbf{A}} = \mathbf{M} \bar{\mathbf{W}}, \bar{\mathbf{b}} = \mathbf{M}\bar{\mathbf{v}}$. If $K(\bar{\mathbf{A}}) \ll K(\bar{\mathbf{W}})$, the CG method solves Eq. (5.3) faster than Eq. (5.1). According to the characteristics of the coefficient matrix of the HASM.MOD equations, the incomplete Cholesky CG (ICCG) and the symmetric SOR-PCG (SSORCG) are used in this chapter.

5.1.1   Incomplete Cholesky Conjugate Gradient (ICCG)

The incomplete Cholesky decomposition of $\bar{\mathbf{W}}$ is $\bar{\mathbf{W}} = \mathbf{M} + \mathbf{R} = \mathbf{L}\mathbf{L}^T + \mathbf{R}$, where **L** denotes the lower triangular matrix, so that $\mathbf{M} = \mathbf{L}\mathbf{L}^T$ can be as close as possible to $\bar{\mathbf{W}}$, and $\mathbf{L}$ has the same sparsity as $\bar{\mathbf{W}}$ or other specified sparsity. In complete Cholesky decomposition, triangular decomposition is performed on the coefficient matrix $\bar{\mathbf{W}}$, where $\bar{\mathbf{W}} = \mathbf{L}\mathbf{L}^{\mathrm{T}}$, while in incomplete Cholesky decomposition, triangular decomposition $\mathbf{L}\mathbf{L}^{\mathrm{T}}$ is conducted on matrix $\bar{\mathbf{W}} - \mathbf{R}$. Since matrix $\mathbf{R}$ varies, the sparse structure of $\mathbf{L}$ could be properly controlled in advance; that is, the zero elements in $\mathbf{L}$ can be specified in advance. Moreover, $\mathbf{L}\mathbf{L}^{\mathrm{T}}$ should be as close to $\bar{\mathbf{W}}$ as possible, which overcomes the issue in complete Cholesky decomposition of the sparsity of $\bar{\mathbf{W}}$ often being damaged. In practical computation, $\mathbf{R}$ usually has many zero elements, and other elements in $\mathbf{R}$ are too large. In this chapter, the unfilled Cholesky decomposition algorithm is used; that is, during Cholesky decomposition of $\bar{\mathbf{W}}$, the non-zero elements are not introduced to the position of zero elements of matrix $\bar{\mathbf{W}}$. The algorithm is given as follows (Golub and Loan 2009):

$$\bar{\mathbf{W}}(k,k) = \sqrt{\bar{\mathbf{W}}(k,k)}$$

for $i = k + 1 : n$
    if $\bar{\mathbf{W}}(i,k) \neq 0$
    $\bar{\mathbf{W}}(i,k) = \bar{\mathbf{W}}(i,k)/\bar{\mathbf{W}}(k,k)$
    end
end
for $j = k + 1 : n$
    for $i = j : n$
        if $\bar{\mathbf{W}}(i,j) \neq 0$
    $\bar{\mathbf{W}}(i,j) = \bar{\mathbf{W}}(i,j) - \bar{\mathbf{W}}(i,k)\bar{\mathbf{W}}(j,k)$
        end
    end
end

With $\mathbf{M} = \mathbf{L}\mathbf{L}^{\mathrm{T}}$ as the preconditioning operator, it is not difficult to verify that the coefficient matrix $\bar{\mathbf{A}} \approx \mathbf{I}$ using the CG method. During the implementation process, $\bar{\mathbf{A}}$ is stored in compressed rows, and $\mathbf{M}z_k = \mathbf{r}_k$ is transformed into solving $\mathbf{L}^{\mathrm{T}}z_k = \mathbf{y}_k$ and $\mathbf{L}\mathbf{y}_k = \mathbf{r}_k$. In this way, the computation size is $\mathrm{O}(n^2)$, which is one order of magnitude lower than that of directly solving $\mathbf{M}z_k = \mathbf{r}_k$. The ICCG method is effective for solving small-scale problems. For large-scale problems, the coefficient matrix can be preprocessed using Cholesky decomposition, and the decomposed lower triangular matrix is used as the input parameters of HASM.MOD. The space complexity and time complexity of ICCG is $\mathrm{O}(n^2)$.

5.1.2 Ssymmetric Successive Over Relaxation-Preconditioned Conjugate Gradient (SSORCG)

The computation of HASM.MOD is mainly composed of multiplication and inversion of matrices, and the time consumed by inversion is much longer than that by multiplication. In the ICCG method, the equation $\mathbf{M}\mathbf{z}_k = \mathbf{r}_k$ needs to be solved in each iteration. To avoid this, the symmetric SOR (SSOR) preconditioned method directly computes the approximate inverse matrix of the coefficient matrix, thereby reducing the computation time. Since this involves only the product of a matrix and vector, parallel computing can be easily implemented in this method.

In the SSOR (Evans and Forrington 1963), coefficient matrix $\bar{\mathbf{W}}$ is assumed to be decomposed into $\bar{\mathbf{W}} = \mathbf{L} + \mathbf{D} + \mathbf{L}^{\mathrm{T}}$, where $\mathbf{D}$ denotes a diagonal matrix composed of diagonal elements of $\bar{\mathbf{W}}$ and $\mathbf{L}$ is a lower triangular matrix composed of the lower triangular elements of $\bar{\mathbf{W}}$. The SSOR preconditioning operator is defined as follows:

$$\mathbf{M} = \mathbf{K}\mathbf{K}^{\mathrm{T}}, \mathbf{K} = \frac{1}{\sqrt{2-w}}\left(\frac{1}{w}\mathbf{D} + \mathbf{L}\right)\left(\frac{1}{w}\mathbf{D}\right)^{\frac{-1}{2}}, 0 < w < 2,$$

The inverse of $\mathbf{K}$ is expressed as: $\mathbf{K}^{-1} = \sqrt{2-w}\left(\frac{1}{w}\mathbf{D}\right)^{\frac{1}{2}}\left(\mathbf{I} + w\mathbf{D}^{-1}\mathbf{L}\right)^{-1}$ $\frac{1}{w}\mathbf{D}^{-1}, \frac{1}{w}\mathbf{D} = \bar{\mathbf{D}},$

Because $\left(\mathbf{I} + \bar{\mathbf{D}}^{-1}\mathbf{L}\right)^{-1} = \mathbf{I} - \bar{\mathbf{D}}^{-1}\mathbf{L} + \overline{(\mathbf{D}}^{-1}\mathbf{L})^2 - \cdots$, $\mathbf{K}^{-1} \approx$ $\sqrt{2-w}\bar{\mathbf{D}}^{\frac{-1}{2}}\left(\mathbf{I} - \bar{\mathbf{D}}^{-1}\mathbf{L}\right)\bar{\mathbf{D}}^{-1} = \sqrt{2-w}\bar{\mathbf{D}}^{\frac{-1}{2}}\left(\mathbf{I} - \mathbf{L}\bar{\mathbf{D}}^{-1}\right) \equiv \bar{\mathbf{K}}.$

Then, the approximate inverse of matrix $\bar{\mathbf{W}}$ is $\bar{\mathbf{M}} = \bar{\mathbf{K}}^{\mathrm{T}}\bar{\mathbf{K}}$. Thus, the problem of solving $\mathbf{M}\mathbf{z}_k = \mathbf{r}_k$ is transformed into solving $\mathbf{z}_k = \bar{\mathbf{M}}\mathbf{r}_k$ in the PCG method, which can be further decomposed into solving $\mathbf{y}_k = \bar{\mathbf{K}}\mathbf{r}_k$ and $\mathbf{z}_k = \bar{\mathbf{K}}^{\mathrm{T}}\mathbf{y}_k$. In this process, the inversion operation is avoided, and the elements in the specific position of $\bar{\mathbf{K}}$ can be expressed explicitly, so the storage capacity is the same as that of the CG method, with the time and space complexities being $O(n^2)$ and $O(n)$, respectively.

5.1.3 Numerical Simulation

Through numerical simulation, the convergence rates of the above-mentioned different PCG methods for solving the HASM.MOD equations are compared to determine the method with the highest accuracy given the same computation time and the least time-consuming method with the same accuracy, thereby providing users with different methods based on their actual needs.

Taking the Gaussian surface as an example, the differences among the CG methods based on diagonal preconditioning (DCG), ICCG and SSORCG are compared. The Gaussian surface is expressed as follows:

$$f(x, y) = 3(1-x)^2 e^{-x^2-(y+1)^2} - 10\left(\frac{x}{5} - x^3 - y^5\right)e^{-x^2-y^2} - \frac{e^{-(x+1)^2-y^2}}{3}$$

The computational domain is $[-3, 3] \times [-3, 3]$, and the range is $-6.5510 < f(x, y) < 8.1062$.

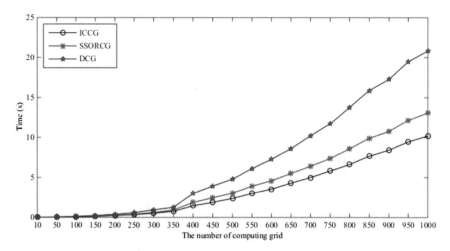

Fig. 5.2 Comparison of the computation times of different preconditioning methods

The iteration convergence criterion for the HASM.MOD equations is set as $\|\mathbf{r}\|_2 <$
10^{-12}, where $\mathbf{r} = \bar{\mathbf{v}} - \bar{\mathbf{W}}x$. Figure 5.2 shows the changes in the computation times
of different methods with the same computational burden and the same convergence
criteria with different computing grids. There is little difference between the three
methods regarding the computation time when the number of grids is less than
350. Therefore, DCG or SSORCG can be used since these methods take up less
memory. As the computational burden increases, the differences among the three
methods become obvious. The computation time of DCG increases significantly as
the number of computing grids increases. The computation time of ICCG changes
slowly, and it is the least time-consuming method, followed by SSORCG. Thus, in
the case of sufficient memory, ICCG can be selected to solve the algebraic equations
in HASM.MOD. The space complexity of SSORCG is $O(n)$, which is one order
of magnitude smaller than that of ICCG, so SSORCG is an ideal method when
considering both memory and computation time.

Since the aim of preconditioning is to reduce the conditional number of the coefficient matrix, the conditional numbers of the coefficient matrices of the equations
preconditioned by different methods are compared. The experiment results show that
for the HASM.MOD matrices of different orders preconditioned by the same method,
the differences in the condition number are less than 1. For matrices of different sizes,
ICCG gives a conditional number of 89, SSORCG gives a conditional number of
180, and DCG gives a conditional number of 756. The results suggest that ICCG has
the fastest the convergence speed, followed by SSORCG.

The computational accuracies of different methods under the same computation
time are compared, and Table 5.1 shows the computation results with 5 outer iterations
and a 101×101 grid.

Table 5.2 shows the computation results with different numbers of outer iterations
when the number of inner iterations is 10 and the number of grids is 501×501.

Table 5.1 Simulation accuracy of different methods with different numbers of inner iterations

Number of inner iterations	ICCG	SSORCG	DCG
5	2.3482	7.4335	7.9532
10	0.0395	0.8129	1.9689
20	0.000026	0.0145	0.0876
50	0.0	0.0	0.000044

Table 5.2 Simulation accuracies of different methods with different numbers of outer iterations

Number of outer iterations	ICCG	SSORCG	DCG
2	0.001163	0.017035	0.094594
4	0.000138	0.002427	0.007262
6	0.000064	0.001169	0.003271
8	0.000042	0.000821	0.002054

The simulation process of HASM.MOD has inner and outer iterations. As shown in Table 5.1, given a fixed number of outer iterations, the computational accuracy of all three methods increases as the number of inner iterations increases. ICCG can achieve relatively high accuracy when the number of inner iterations is 10, and the computational accuracy improves rapidly as the number of inner iterations increases. DCG has the lowest accuracy. In Table 5.2, given the same number of inner iterations, the simulation accuracy of different methods increases with increasing the number of outer iterations, yet the computational accuracy of each method increases slowly. Of the three methods, ICCG has the best performance. The results show that compared with outer iteration, inner iteration plays a more important role in improving the simulation accuracy of HASM.MOD. Moreover, with the same number of iterations, ICCG has the highest accuracy, followed by SSORCG, and DCG has the lowest accuracy. SSORCG is a good option when memory is limited.

5.2   Parallelization of HASM.MOD

The PCG method for solving the HASM.MOD equations is eventually transformed into solving the product of a sparse matrix and vector and the inner product of vectors. The matrix and vector can be decomposed into several row or column elements; the process of multiplying the matrix and vector can be transformed into the process of first multiplying these row or column elements and then summing all multiplication results and so can the multiplication of vectors. According to certain rules, the elements of the vector and matrix can be decomposed into various processes that are not highly related to each other, so it is easy to realize parallel computing. In this section, the realization of the parallelization of HASM.MOD is explored using an IBM blade-cluster, open Linux system, MPI parallel Environment, and PCG.

For serial implementation of HASM.MOD, the formation stage of the right end term of the HASM.MOD equations occupies the most memory. Assuming that the size of the computational grid is m × n, the memory required at this stage is approximately $19 \times m \times n \times 8$ bytes. The process of solving the equations takes up the most time (approximately 98%) in the whole computation process of HASM.MOD, and the formation of the preconditioned matrix is the second most time-consuming. In comparison, the time consumption during the establishment of sampling equation and the input and output of data can be ignored.

The parallelization of the PCG method aims to distribute data to different processes according to certain methods and to update the data in each process to improve the computation speed. The matrix preconditioned by DCG often has a large conditional number, which results in slow convergence. ICCG is a better preconditioning method on serial computers, but its inherent forward–backward dependency is not conducive to parallelization. SSORCG can be easily parallelized because the inverse matrix of the preconditioned matrix is formed only by the product, addition and subtraction of the matrix and vector.

The analysis of the implementation process of the PCG method shows that this method is eventually transformed into solving the product of a sparse matrix and vector, the inner product of vectors, the product of a number and vector, and the summing and subtraction of vectors. These operations do not depend on the order of elements, which is ideal for parallelization. The code for parallelization of the PCG method is as follows:

Algorithm 3: Preconditioned Conjugate Gradient (PCG) parallel version:

$$
\begin{aligned}
&\text{Given an initial } \boldsymbol{x}_0, \\
&\quad \text{Scatter } \boldsymbol{x}_0, \overline{\boldsymbol{v}} \text{ in the processes,} \\
&k = 0, \boldsymbol{r}_0 = \overline{\boldsymbol{v}} - \overline{\boldsymbol{W}} \boldsymbol{x}_0, \text{ (parallel)} \\
&\quad \text{while } (\boldsymbol{r}_k \neq 0) \\
&\qquad\qquad \text{solve } \boldsymbol{M} \boldsymbol{z}_k = \boldsymbol{r}_k \text{ (parallel)} \\
&\qquad\qquad k = k+1 \\
&\qquad\qquad \text{If } k = 1 \\
&\boldsymbol{p}_1 = \boldsymbol{z}_0 \\
&\qquad\qquad\quad \text{else} \\
&\qquad\quad \beta_k = \boldsymbol{r}_{k-1}^T \boldsymbol{z}_{k-1} / \boldsymbol{r}_{k-2}^T \boldsymbol{z}_{k-2} \text{ (parallel)} \\
&\quad \boldsymbol{p}_k = \boldsymbol{z}_{k-1} + \beta_k \boldsymbol{p}_{k-1} \text{ (parallel)} \\
&\qquad\qquad\quad \text{end} \\
&\qquad\quad \alpha_k = \boldsymbol{r}_{k-1}^T \boldsymbol{z}_{k-1} / \boldsymbol{p}_k^T \overline{\boldsymbol{W}} \boldsymbol{p}_k \text{ (parallel)} \\
&\qquad\quad \boldsymbol{x}_k = \boldsymbol{x}_{k-1} + \alpha_k \boldsymbol{p}_k \text{ (parallel)} \\
&\qquad\quad \boldsymbol{r}_k = \boldsymbol{r}_{k-1} - \alpha_k \overline{\boldsymbol{W}} \boldsymbol{p}_k \text{ (parallel)} \\
&\qquad\qquad \text{end} \\
&\boldsymbol{x} = \boldsymbol{x}_{k-1}
\end{aligned}
$$

The computational burden is the initial computational burden + the number of iterations × the computational burden in each iteration; that is,

$2 \times$ dot_product_flops $+ 1 \times$ psolve_flops $+$ iter_number $\times(2 \times$ dot_product_flops $+ 1 \times$ matvec_flops $+ 1 \times$ psolve_flops $+ 3 \times$ axpy_flops $+ 2)$, where dot_product_flops is the inner product of the vectors, the computational burden is $2 \times N$, and N is the length of the vector; psolve_flops refers to $z_k = M^{-1}r_k$, and the computation burden is determined by the selection of the preconditioned matrix; matvec_flops denotes the product of the sparse matrix and vector, and the computational burden is $2 \times$ the number of non-zero elements; axpy_flops represents the product of the constant and vector and the addition and subtraction of vectors, and the computational burden is $2 \times N$; and 2 represents two division operations.

In this chapter, sparse matrices are stored by means of compressed row storage (CSR). In CSR, the value of each non-zero element of sparse matrix **A**, the column of each non-zero element, and the index of the first non-zero element in each row are stored; i.e., there are a total of three arrays, **a**, **ja**, **ia** ($A \in R^{n \times n}$, and nz denotes the number of non-zero elements):

a(nz) records the value of each non-zero element.
ja(nz) records the column of each non-zero element.
ia(n + 1) records the index of the first non-zero element in each row in arrays **a**(nz) and **ja**(nz), where **ia**(n + 1) = nz + 1.

In parallel computing, sending and receiving messages results in large system overhead (Du 2001). Although the default method for storing matrices in Fortran is column storage, matrix vector multiplication that relies on column storage can result in a large amount of communication between processes. Hence, for the multiplication of the sparse matrix and vector, according to the storage structure of the sparse matrix, the pseudocode is as follows:

```
MPI_ALLGATHERV(x0(istart), jlen(iproc), MPI_REAL8, xx, jlen, jdisp, &
                &MPI_REAL8, icomm, ierr)
do i=1, n
 sum=0
 do k=ia(i), ia(i+1)-1
  sum=sum+a(k)*x0(ja(k))
 enddo
 xx(i)=sum
 enddo
```

In this method, only the operation that $\times 0$ performs group collection from other processors through MPI_ALLGATHERV before each computation must be guaranteed so that $\times 0$ is an updated vector for each iteration.

The pseudocode of the inner product of the vectors is as follows:

```
do i = istart, iend

 sum = sum + v(i)*w(i)

enddo
```

call MPI_ALLREDUCE(sum, s, 1, MPI_REAL8, MPI_SUM, root, icomm, ierr)

where istart and iend are the start and end positions of vectors v and w in each process. After each process, the results are sent to each process according to the specified operation MPI_SUM through the reduction of MPI_ALLREDUCE.

For the multiplication of numbers and vectors and the addition and subtraction of vectors, the pseudocode is as follows:

```
do i=istart, iend
    jg(i)=v(i)+alpa*w(i)
enddo
do irank=0, nproc-1 ! nproc is the total number of processes
    jlen(irank)=iend-istart+1! jlen denotes the number of data points in the
message-sending buffer zone
    jdisp(irank)=istart-1! jdisp is the offset of the received data
enddo
call MPI_ALLGATHERV(jg(istart), jlen(iproc), MPI_REAL8, xx, jlen, jdisp, &
                    &MPI_REAL8, icomm, ierr)
```

The execution time required for parallel computing consists of three parts, i.e., the computation time, communication time, and parallel overhead time. The parallel overhead time mainly refers to the time of parallel management, grouping operation, and query operation by the operating system. The communication time is mainly the delay needed for information and data transmission between processors. The speed-up ratio is an important criterion for evaluating parallel algorithms. S_p is the speed-up ratio of p processors and can be defined as follows (Xie et al. 2003):

S_p = time to solve the problem on a single machine/time for p processors to solve the problem.

Taking the Gaussian surface as an example, the efficiency of the HASM.MOD parallel algorithm is analysed with a computational domain of $[-3,3] \times [-3,3]$ and a grid size of 2000×2000. Figure 5.3 shows the real (a) and simulated (b) Gaussian surfaces. During the computation, the number of processors gradually increases from 2. When there are 2 processors, only one processor participates in the computation, which is equivalent to serial execution. Figure 5.4 shows the trends of the computation time and speed-up ratio of the parallel algorithm with increasing number of processors. As the number of processors increases, the computation time significantly decreases. As shown in Fig. 5.4b, when the number of processors reaches 10, the speed-up ratio is the maximum; as the number of processors continue to increase, the speed-up ratio decreases, and the communication overhead increases.

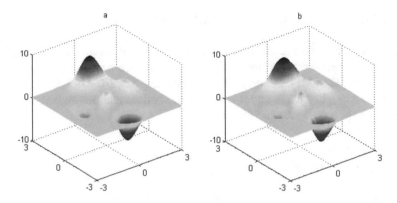

Fig. 5.3 Real (**a**) and simulated (**b**) Gaussian surfaces

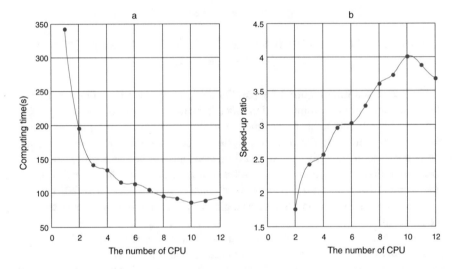

Fig. 5.4 Computation time and speed-up ratio of the HASM.MOD parallel algorithm

5.3   Summary

HASM.MOD is ultimately transformed into the solving of large-scale ill-conditioned sparse linear equations. An efficient solving method is particularly important for HASM.MOD. In this chapter, according to the characteristics of the coefficient matrix of the HASM.MOD equations, on the basis of the CG method, the optimal preconditioning operator is selected to transform the original equations into well-conditioned equivalent linear equations, which accelerates the convergence speed of HASM.MOD. Moreover, parallel computing of HASM.MOD is realized in an MPI parallel environment.

Using the Gaussian surface as the subject, the results of numerical simulation show the following: (1) When the computational burden is small, there are small differences among the three preprocessing methods of DCG, SSORCG and ICCG in terms of computation time. Therefore, DCG or SSORCG can be used since they take up less memory. When the computational burden is large, there are obvious differences among the three methods. The computation time of DCG increases significantly as the number of computing grids increases. ICCG yields the shortest computation time, followed by SSORCG. Therefore, with sufficient memory, ICCG can be used to solve the algebraic equations in HASM.MOD. When both memory and computation time are considered, SSORCG is an ideal method. (2) Given the same numbers of inner and outer iterations, ICCG has the highest accuracy, followed by SSORCG, and DCG has the lowest accuracy. (3) Inner iteration has a greater influence on the results of HASM.MOD than outer iteration. (4) In the parallelization of HASM.MOD, as the number of processors increases, the computation time decreases significantly. When the number of processors reaches a certain value, the speed-up ratio reaches its maximum. After the maximum, the speed-up ratio decreases with increasing number of processors. This is because the decrease in the computation time with an increasing number of processors cannot offset the communication overhead caused by an increasing number of processors.

References

Du ZH. 2001. Parallel Programming Techniques for High Performance Computing-MPI Progtamming. Beijing: Tsinghua University Press.

Evans DJ, Forrington CVD. 1963. An iterative process for optimizing symmetric successive over-relaxation. The Computer Journal, 6(3): 271–273.

Golub GH, Van Loan CF. 2009. Matrix Computations. Posts & Telecom Press.

Xie C, Mai LD, Dou ZH, et al. 2003. Research and analysis of parallel computing system speedup. Computer Engineering and Applications, 39: 66–68

Chapter 6
High Accuracy Surface Modelling of Average Seasonal Precipitation in China Over a Recent Period of 60 Years

Based on the monthly precipitation data collected at 711 meteorological stations in China from 1951 to 2010, the average seasonal precipitation from 1951 to 2010 is simulated in different areas of China according to the agro climatic type. According to the precipitation characteristics, the geographical, topographic and local topographic factors influencing the precipitation in different subareas are explored, and trend fitting of precipitation in each subarea is carried out via polynomial regression and stepwise regression. Then, the improved high-accuracy surface modelling (HASM) method (HASM.MOD) is used to iteratively correct the residual after removing the trend in each simulated area, and the performance of the method is verified. Moreover, to ensure simulation accuracy at the boundary, a buffer zone is set for each subarea according to the distance between the meteorological stations within the subarea, thereby extending the HASM interpolation area to the buffer zone. The simulation results show that HASM had higher accuracy than classical interpolation methods in different subareas and across different seasons. The above method is applied to analyse the distribution characteristics of precipitation in different subareas during the same season and to simulate the spatial distribution of average precipitation from 1951 to 2010 in different seasons. The results are consistent with the actual distribution characteristics of precipitation in China.

6.1 Introduction

Precipitation, as an important agroclimatic resource, is the main driving variable of terrestrial biosphere models (Yue 2007; Moulin 2009; Cai 2009; Bannayan 2011). However, the number of meteorological stations is limited, and the distribution of meteorological elements is spatially discrete. Extensive studies on obtaining continuous spatial characteristics of climate elements have been carried out worldwide. Yan et al. simulated the climate in China using the AUNSPLIN model based on the principle of spline interpolation. Daly et al. applied the PRISM model to analyse the

© The Author(s), under exclusive license to Springer Nature Singapore Pte Ltd. 2021   125
N. Zhao and T. Yue, *High Accuracy Surface Modeling Method: The Robustness*,
https://doi.org/10.1007/978-981-16-4027-8_6

spatial distribution of meteorological elements in the United States. Zhang et al. used an artificial neural network to analyse the average temperature from 1951 to 2010 in China. Joly et al. (2011) simulated the spatial variations in temperature in France by means of local interpolation. Marquinez et al. combined geographical and topographic factors to simulate precipitation in mountainous areas using the multivariate statistical method. Since the 1990s, the rapid development of geographic information system (GIS) technology in China has facilitated many spatial interpolation methods, e.g., the inverse distance weighting (IDW) method, kriging method, and spline method. These methods offer a solution for determining the spatial distribution of meteorological elements. However, all these methods feature low interpolation efficiency either because the processing time is long due to model complexity or because the model is too simple to yield high-accuracy results. Moreover, the performance of different spatial methods in interpolating meteorological elements varies from region to region, and there is no absolute optimal interpolation method.

The precipitation in China is regional and seasonal and is greatly influenced by geographical and topographic factors. Considering the influences of longitude, latitude, and altitude, Liu et al. interpolated meteorological elements using 1 km grids and then used the IDW method to simulate the residual value after removing the trend to obtain the corrected value in each grid. Though geographical and topographic factors were considered, the influence of local topographic factors on precipitation was ignored.

Based on the principal theorem of surface theory, HASM was developed in the 1990s (Yue 2010). HASM solves the problems of peak-cutting and oscillating boundaries in interpolation. Compared with classical interpolation methods, such as the IDW, kriging and spline methods, HASM has remarkably better simulation accuracy and numerical characteristics (Yue et al. 2004, 2006). HASM has been successfully applied to the simulation of population distribution and soil pH value (Tian 2005; Shi 2009). However, in the field of climate, HASM has underperformed.

In light of the above problems, HASM is used in this chapter to simulate the average seasonal precipitation in different subareas of China over a recent period of 60 years on 1 km grids, with the consideration of the geographical and topographic factors in each subarea. Moreover, the local topographic factors influencing precipitation in each subarea, i.e., the average topographic factors within 2–10 km of each grid point, are considered. Then, based on stepwise regression, the precipitation in each subarea is fitted by polynomial regression, and the residual is corrected using HASM after removing the trend. To deal with the problem of low accuracy at the boundary due to a lack of sampling points outside the boundary, a buffer zone is established for each subarea, and the actual computational domain was controlled within the buffer zone, which improved the interpolation accuracy of HASM near the regional boundaries.

6.2 Data

The data used in this chapter are the average monthly precipitation collected at more than 700 meteorological stations in China since 1951 and the longitude and latitude information of the meteorological stations. The meteorological stations are unevenly distributed in China; those in the west regions are extremely sparse compared with those in the east regions (Fig. 6.1). The meteorological stations are set up at different times. Before 1950, there were very few meteorological stations to the west of 100° E in China, and large-scale construction of meteorological stations began in the early 1950s. An analysis of these meteorological stations shows that 35-year observation records are available in 681 stations, and 21–35-year observation records are available in 30 stations, which are located in western China. Considering the scarcity of meteorological stations in western China, data of 711 stations are used in this study, and the average seasonal precipitation collected at each station from 1951 to 2010 is calculated.

China is located in southeast Eurasia near the vast Pacific Ocean (15–55° N, 70–140° E), and the differences in the spatial distribution of precipitation in China are obvious. To clearly reflect the variations in precipitation in various parts of China, according to the distribution of agriculture and animal husbandry and diverse climatic conditions in China (Li 1988), the study area is divided into following four subareas:

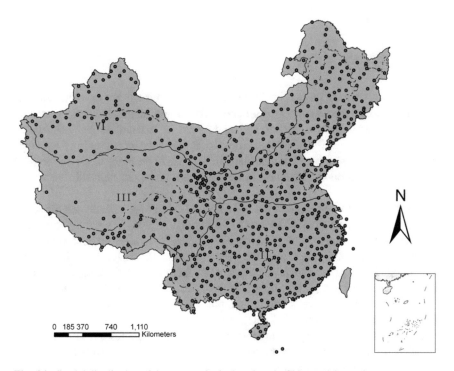

Fig. 6.1 Spatial distribution of the meteorological stations in China and four subareas

Table 6.1 Area and number of stations in each subarea

Subarea	Area (km^2)	Number of meteorological stations	Number of meteorological stations in the computational domain	Percentage of meteorological stations in the computational domain (%)
I	2,083,497.8422	201	283	39.80
II	2,546,959.3791	287	342	48.10
III	2,493,243.1265	110	169	15.47
VI	2,362,054.7859	113	165	15.89

(I) East China to the north of the Qinling Mountains and the Huaihe River, (II) eastern China to the south of the Qinling Mountains and the Huaihe River, (III) the Qinghai-Tibet Plateau, and (VI) the inland northwestern China (Fig. 6.1). To improve the simulation accuracy at the boundary of each subarea, a buffer zone with a radius of 100 km is set for each subarea according to the distance between meteorological stations, and the actual interpolation area refers to each subarea and the surrounding buffer zone. Table 6.1 shows the area and number of stations in each subarea.

6.3   Methods

6.3.1   Polynomial Regression

According to statistics, the spatial distribution of a variable can be regarded as a realization of a spatial process under the influence of multiple factors. This spatial process includes global trends, local effects, and random errors. To simulate the spatial distribution of precipitation, the geographical and topographic factors affecting the precipitation in each subarea should be identified first to present the regression relationships between precipitation and geographical/topographic factors in each subarea during different seasons.

Liu et al. (2004) investigated the influences of longitude, latitude and elevation on precipitation. Daly et al. (2008) and Wotling et al. (2000) pointed out that local topographic factors have strong influences on precipitation. In previous studies on precipitation simulation in China, the influences of local topographic factors are often ignored. The geographical and topographic factors are considered in this study, including the longitude (unit: m), latitude (unit: m), elevation (unit: m), slope (unit: degree), aspect (unit: degree), and topographic relief (unit: m). Table 6.2 shows the local topographic factors that influence precipitation.

The elevation, slope, aspect, topographic relief, and local terrain factors are all extracted from digital elevation model (DEM) raster data using ArcGIS. The relationship between the geographical, topographic factors and precipitation is fitted

Table 6.2 Local topographic factors influencing precipitation

Local topographic factor	Meaning	Local topographic factor	Meaning
DEM25	Average elevation within 2.5 km around the interpolation point	Aspect25	Average aspect within 2.5 km around the interpolation point
DEM50	Average elevation within 5 km around the interpolation point	Aspect50	Average aspect within 5 km around the interpolation point
DEM100	Average elevation within 10 km around the interpolation point	Aspect100	Average aspect within 10 km around the interpolation point
Slope25	Average slope within 2.5 km around the interpolation point	Relief25	Average topographic relief within 2.5 km around the interpolation point
Slope50	Average slope within 5 km around the interpolation point	Relief50	Average topographic relief within 5 km around the interpolation point
Slope100	Average slope within 10 km around the interpolation point	Relief100	Average topographic relief within 10 km around the interpolation point

using Eq. (6.1):

$$
\begin{aligned}
\mathrm{Pre} = {}& \mathbf{b}_0 + \mathbf{b}_1\mathbf{x}_1 + \mathbf{b}_2\mathbf{x}_2 + \mathbf{b}_3\mathbf{x}_3 + \mathbf{b}_4\mathbf{x}_4 + \mathbf{b}_5\mathbf{x}_5 + \mathbf{b}_6\mathbf{x}_6 + \cdots + \mathbf{b}_{11}\mathbf{x}_1^2 \\
& + \mathbf{b}_{22}\mathbf{x}_2^2 + \mathbf{b}_{33}\mathbf{x}_3^2 + \mathbf{b}_{44}\mathbf{x}_4^2 + \mathbf{b}_{55}\mathbf{x}_5^2 + \mathbf{b}_{66}\mathbf{x}_6^2 + \mathbf{b}_{12}\mathbf{x}_1\mathbf{x}_2 + \mathbf{b}_{13}\mathbf{x}_1\mathbf{x}_3 \\
& + \mathbf{b}_{14}\mathbf{x}_1\mathbf{x}_4 + \mathbf{b}_{15}\mathbf{x}_1\mathbf{x}_5 + \mathbf{b}_{16}\mathbf{x}_1\mathbf{x}_6 + \mathbf{b}_{23}\mathbf{x}_2\mathbf{x}_3 + \cdots,
\end{aligned}
\tag{6.1}
$$

where the linear terms in Eq. (6.1) refer to the geographical and topographic factors, including the local factors, and the quadratic and cross terms refer to the square of each factor and the product of any two factors, respectively. Since not all factors have significant influences on precipitation, there can be multiple collinearities among some factors. Stepwise regression (Deng 2006) is used to determine the optimal subset from all possible combinations of these factors. Based on the fitting relationship obtained by stepwise regression, the geographical/topographic factors in the relationship equation are all significantly correlated with precipitation, and the quadratic sum of the deviation between the fitted value and the real value can be minimized.

Polynomial regression and stepwise regression are jointly applied to each interpolation area to obtain the geographical/topographic factors influencing precipitation in each subarea during different seasons as well as the coefficient of determination R^2 of the corresponding regression equation (Table 6.3). Table 6.3 shows the regional

Table 6.3 Geographical and topographic factors influencing precipitation in each subarea during different seasons and the R^2 of the corresponding regression equation

Season	Subarea	Geographical/topographic factors	R^2	Adj_ R^2
Spring	I	DEM, Lat, Aspect, Relief100, Lat2, Relief502, DEM*Lat,DEM*Relief, Lon*Lat, Aspect*Relief	0.8512	0.8444
	II	DEM100, Lon, Lat, Aspcet100, Relief50, DEM252, Lat2, Aspect252, Slope1002, Lon*Lat	0.7600	0.7522
	III	Lon, Lat, Relief252, Lon*Lat	0.5697	0.5593
	VI	Aspect, Relief100, DEM2, Lon2, Aspect502, Relief1002, DEM*Lon, DEM*Aspect, Lat*Slope	0.5752	0.5505
Summer	I	DEM100, Lon, Lat, Relief25, Relief50, DEM2, Lon2, Lat2, Slope502, Relief502, DEM*Lon, Lon*Lat	0.7906	0.7811
	II	DEM50, DEM100, Lon, Lat, Slope100, Relief, Relief100, DEM502, Lon2, Relief2, Relief502, DEM*Lon, Lon*Lat, Aspect*Slope	0.5664	0.5463
	III	DEM, DEM25, DEM100, Lat, Relief, Relief100, DEM*Lat DEM1002, Lon2, Lat2, Relief252, Relief502, DEM*Lon	0.8825	0.8701
	VI	Lon, Lat, Relief100, DEM252, Lon2, Lat2, Aspect1002, Relief1002, DEM*Lon, Lon*Lat	0.9009	0.8944
Autumn	I	Lat, Relief100, DEM2, DEM1002, Lat2, Relief1002, DEM*Lon, DEM*Relief, Lon*Lat	0.8226	0.8165
	II	DEM100, Lon, Lat, Lon2, Lat2, Aspect2, Slope502, Slope1002, Relief252, DEM*Lon, Lon*Relief	0.4809	0.4605
	III	Lon, Lat, Aspect502, Lon*Lat, Lon*Relief, Aspect*Relief	0.7828	0.7748
	VI	Lon, Lat, Relief100, Lon2, Lat2, Aspect1002, Relief1002, DEM*Relief, Lon*Lat, Lon*Relief	0.8394	0.8289
Winter	I	DEM, Lon, Lat, DEM1002, Lon2, Lat2, DEM*Lon, DEM*Lat, Lon*Lat	0.8817	0.8777
	II	Lon, Lat, Aspect100, Slope100, Lon2, Lat2	0.7553	0.7506
	III	Lon, Lat, Relief252, DEM*Relief, Lon*Lat, Lon*Relief	0.4014	0.3715
	VI	Lon, Lat, DEM2, DEM1002, Lon2, Lat2, Slope1002, Lon*Lat, Aspect*Relief	0.5906	0.5640

DEM: elevation, Lat: latitude, Lon: longitude, Aspect: aspect, Slope: slope, Relief: topographic relief

and seasonal characteristics of precipitation in China.

6.3.2 HASM

If the equation of a surface is $\mathbf{z} = \mathbf{f}(\mathbf{x}, \mathbf{y})$, then HASM can be reduced to solving the differential equation set:

$$\begin{cases} \mathbf{f}_{xx} = \Gamma_{11}^1 \mathbf{f}_x + \Gamma_{11}^2 \mathbf{f}_y + \mathbf{L}(\mathbf{EG} - \mathbf{F}^2)^{-1/2} \\ \mathbf{f}_{yy} = \Gamma_{22}^1 \mathbf{f}_x + \Gamma_{22}^2 \mathbf{f}_y + \mathbf{N}(\mathbf{EG} - \mathbf{F}^2)^{-1/2} \end{cases} \tag{6.2}$$

where $\mathbf{E} = 1 + \mathbf{f}_x^2$, $\mathbf{F} = \mathbf{f}_x\mathbf{f}_y$, $\mathbf{G} = 1 + \mathbf{f}_y^2$, $\mathbf{L} = \dfrac{\mathbf{f}_{xx}}{\sqrt{1+\mathbf{f}_x^2+\mathbf{f}_y^2}}$, $\mathbf{N} = \dfrac{\mathbf{f}_{yy}}{\sqrt{1+\mathbf{f}_x^2+\mathbf{f}_y^2}}$,

$$\Gamma_{11}^1 = \frac{1}{2}(\mathbf{GE}_x - 2\mathbf{FF}_x + \mathbf{FE}_y)(\mathbf{EG} - \mathbf{F}^2)^{-1}, \; \Gamma_{11}^2$$
$$= \frac{1}{2}(2\mathbf{EF}_x - \mathbf{EE}_y - \mathbf{FE}_x)(\mathbf{EG} - \mathbf{F}^2)^{-1},$$

$\Gamma_{22}^1 = \frac{1}{2}(2\mathbf{GF}_y - \mathbf{GG}_x - \mathbf{FG}_y)(\mathbf{EG} - \mathbf{F}^2)^{-1}, \; \Gamma_{22}^2 = \frac{1}{2}(\mathbf{EG}_y - 2\mathbf{FF}_y + \mathbf{FG}_x)(\mathbf{EG} - \mathbf{F}^2)^{-1}.$

Using the finite difference method mbined with constraint control of sampling points, Equation set (6.2) is discretized into a constrained least square problem (Shi et al. 2009):

$$\begin{cases} \min \left\| \begin{bmatrix} \mathbf{A} \\ \mathbf{B} \end{bmatrix} \mathbf{z}^{(n+1)} - \begin{bmatrix} \mathbf{d}^{(n)} \\ \mathbf{q}^{(n)} \end{bmatrix} \right\| \\ \text{s.t.} \quad \mathbf{S}\mathbf{z}^{(n+1)} = \mathbf{k} \end{cases}_2 \tag{6.3}$$

where A, B, $\mathbf{d}^{(n)}$ and $\mathbf{d}^{(n)}$ are the coefficient matrices and the right end terms corresponding to the discretized differential equations and S and k refer to the matrix formed by the sampling points and the column vector composed of the corresponding sampling values, respectively. If $(\mathbf{x}_i, \mathbf{y}_j, \bar{\mathbf{f}}_{ij})$ is the coordinate and sampling value of the t-th sampling point, then $\mathbf{S}(\mathbf{t}, (\mathbf{i}-1) \times \mathbf{N} + \mathbf{j}) = 1$, $\mathbf{k}(\mathbf{t}, (\mathbf{i}-1) \times \mathbf{N} + \mathbf{j}) = \bar{\mathbf{f}}_{ij}$, and N denotes the number of columns of matrix S.

The Lagrangian multiplier λ is introduced, and the least square problem (6.3) is transformed into the following algebraic equations:

$$\mathbf{Wz} = \tilde{\mathbf{v}} \tag{6.4}$$

$$\text{where } \mathbf{W} = \begin{bmatrix} \mathbf{A}^T & \mathbf{B}^T & \lambda S^T \end{bmatrix} \begin{bmatrix} \mathbf{A} \\ \mathbf{B} \\ \lambda S \end{bmatrix}, \ \tilde{v} = \begin{bmatrix} \mathbf{A}^T & \mathbf{B}^T & \lambda S^T \end{bmatrix} \begin{bmatrix} \tilde{\mathbf{d}} \\ \tilde{\mathbf{q}} \\ \lambda k \end{bmatrix}.$$

The coefficient matrix in Eq. (6.4) is a symmetric positive definite large sparse matrix. The CG method is a well-known solution to this problem (Zhang and Liao 2011). However, due to the characteristics of the coefficient matrix of the equations, the convergence rate is rather slow when solving Eq. (6.4) using the CG method. In view of this, the symmetric successive over-relaxation preconditioned conjugate gradient (SSORCG) method is used, which not only increases the convergence rate but also reduces the memory overhead during calculation. Moreover, parallel computing can be applied with this method.

With the SSORCG method, the coefficient matrix is set as follows: $\mathbf{W} = \mathbf{L} + \mathbf{D} + \mathbf{L}^T$, where $\mathbf{D}$ denotes the diagonal matrix composed of the diagonal elements of W and L is the lower triangular matrix formed by the lower triangular part of W. The symmetric successive over-relaxation preconditioning operator is defined as follows: where

$$\mathbf{M} = \mathbf{K}\mathbf{K}^T, \ \text{where } \mathbf{K} = \frac{1}{\sqrt{2-\mathbf{w}}}\left(\frac{1}{\mathbf{w}}\mathbf{D} + \mathbf{L}\right)\left(\frac{1}{\mathbf{w}}\mathbf{D}\right)^{-1/2}, 0 < \mathbf{w} < 2, \quad (6.5)$$

Then, $\mathbf{K}^{-1} = \sqrt{2-\mathbf{w}}\left(\frac{1}{\mathbf{w}}\mathbf{D}\right)^{\frac{1}{2}}\left(\mathbf{I} + \mathbf{w}\mathbf{D}^{-1}\mathbf{L}\right)^{-1}\frac{1}{\mathbf{w}}\mathbf{D}^{-1}$. Let $\frac{1}{\mathbf{w}}\mathbf{D} = \bar{\mathbf{D}}$, and since $\left(\mathbf{I} + \bar{\mathbf{D}}^{-1}\mathbf{L}\right)^{-1} = \mathbf{I} - \bar{\mathbf{D}}^{-1}\mathbf{L} + \overline{(\mathbf{D}}^{-1}\mathbf{L})^2 - \cdots$,

$\mathbf{K}^{-1} \approx \sqrt{2-\mathbf{w}}\bar{\mathbf{D}}^{\frac{1}{2}}\left(\mathbf{I} - \bar{\mathbf{D}}^{-1}\mathbf{L}\right)\bar{\mathbf{D}}^{-1} = \sqrt{2-\mathbf{w}}\bar{\mathbf{D}}^{\frac{-1}{2}}\left(\mathbf{I} - \mathbf{L}\bar{\mathbf{D}}^{-1}\right) \equiv \bar{\mathbf{K}}$. Let $\bar{\mathbf{M}} = \bar{\mathbf{K}}^T\bar{\mathbf{K}}$, then $\bar{\mathbf{M}}$ can be regarded as the preconditioning operator of Eq. (6.4).

When solving Eq. (6.4) using the SSORCG method, HASM can achieve the present simulation accuracy through control of the iterative steps. In addition, the simulation accuracy of HASM can be adjusted based on actual needs.

6.4 Model Verification and Result Analysis

6.4.1 Model Verification

Classical interpolation methods including kriging, IDW and spline have been embedded in many types of software and been successfully applied in various fields (Kleijen 2005; Sinoweki 1997; Schoonmakers 2003). In this study, a comparative analysis of HASM, kriging, IDW and spline is carried out to verify the accuracy of HASM. Based on a statistical analysis of the average precipitation data collected at 711 meteorological stations in China from 1951 to 2010, 20% of the data are randomly selected as validation data and 80% are used as test data for interpolation by the HASM, kriging, IDW and spline methods. Taking the data in summer and winter as examples, the root mean square error (RMSE), mean absolute error (MAE)

Table 6.4 RMSE, MAE, MRE for HASM, kriging, IDW and spline results (summer)

Subarea	Method	RMSE	MAE	MRE
I	HASM	32.9178	21.5076	0.0625
	Kriging	38.5382	28.0876	0.0893
	IDW	44.3066	30.7472	0.0986
	Spline	65.1995	40.7834	0.1237
II	HASM	103.8364	64.3761	0.0952
	Kriging	106.6764	69.9295	0.1143
	IDW	108.5141	72.7767	0.1199
	Spline	157.7672	102.3794	0.1700
III	HASM	48.0262	38.2949	0.2380
	Kriging	60.5899	42.9434	0.2159
	IDW	61.5819	44.9529	0.2547
	Spline	100.3280	73.6587	0.4199
VI	HASM	29.0980	20.7774	0.2637
	Kriging	29.1699	21.4244	0.2665
	IDW	34.8156	27.3340	0.2902
	Spline	38.3273	27.2715	0.3339

and mean relative error (MRE) of the different methods in different simulated areas are calculated. The equations are as follows:

$$\text{RMSE} = \sqrt{\frac{\sum_{i=1}^{N}\left(f_{ij} - \tilde{f}_{ij}\right)^2}{N}}, \ \mathbf{MAE} = \frac{\left|f_{i,j} - \tilde{f}_{i,j}\right|}{N}, \ \mathbf{MRE} = \frac{\left|f_{i,j} - \tilde{f}_{i,j}\right|}{f_{ij}N},$$

where $f_{i,j}$ denotes the measurement value at each meteorological station, $\tilde{f}_{i,j}$ represents the simulated value, and N refers to the number of stations. Tables 6.4 and 6.5 show the simulation results.

As shown in Tables 6.4 and 6.5, the interpolation results of HASM are superior to those of other three classical interpolation methods in different simulation areas. In all areas, HASM shows great interpolation accuracy in different seasons. Moreover, HASM has the highest interpolation accuracy. The RMSE value of the kriging method is smaller than that of IDW and spline methods because the interpolation surface of the kriging method is relatively smooth. The maximum simulated value obtained by the spline method is often greater than the actual value, and the minimum simulated value is smaller than the actual value. Thus, the spline method has the highest RMSE value of the three classical methods and the lowest accuracy.

Table 6.5 RMSE, MAE, MRE for HASM, kriging, IDW and spline results (winter)

Subarea	Methods	RMSE	MAE	MRE
I	HASM	4.6071	3.0086	0.3970
	Kriging	4.6182	3.0299	0.2455
	IDW	4.8281	3.0404	0.2441
	Spline	8.2899	4.8396	0.3986
II	HASM	16.6755	11.6754	0.2484
	Kriging	18.6231	12.0001	0.2275
	IDW	19.6859	13.0721	0.2984
	Spline	25.6599	16.9093	0.5046
III	HASM	7.7879	3.5947	1.0458
	Kriging	12.9323	6.4224	2.4672
	IDW	12.8041	6.0721	1.6086
	Spline	15.9087	8.2058	2.2777
VI	HASM	8.1837	3.7109	0.6987
	Kriging	8.3687	3.6779	0.4463
	IDW	8.6174	3.8821	0.5527
	Spline	9.2759	4.9617	0.9485

6.4.2 Results

The 1951–2010 average seasonal precipitation in different subareas is simulated by polynomial regression and HASM residual interpolation, and the results are shown in Figs. 6.2 and 6.3. Figure 6.2 shows the average seasonal precipitation from 1951 to 2010 for each subarea. In each subarea, the average precipitation in each grid is computed by HASM interpolation for 1951–2010, and the average precipitation in the subarea is the arithmetic average of the precipitation in all grids within the area. The precipitation modified by HASM can accurately reflect the actual distribution of precipitation in the simulated area; thus, the regional average precipitation exhibits high accuracy.

Figure 6.3 shows the spatial distribution of the average seasonal precipitation in China.

Through the simulated average seasonal precipitation, the following conclusions can be drawn using the statistical analysis function of ArcGIS.

In spring, the precipitation in China is mainly concentrated in the southeast coastal areas, and there is generally more precipitation in the south wing of the East Himalayas and the northwest of the Kunlun Mountains. In eastern China to the north of the Qinling Mountains and the Huaihe River, the average precipitation in spring is approximately 91 mm. The average precipitation in the area to the south of the Qinling Mountains and the Huaihe River is 351.96 mm. More precipitation is often found in the hilly regions and basins to the south of the Yangtze River, Zhejiang and

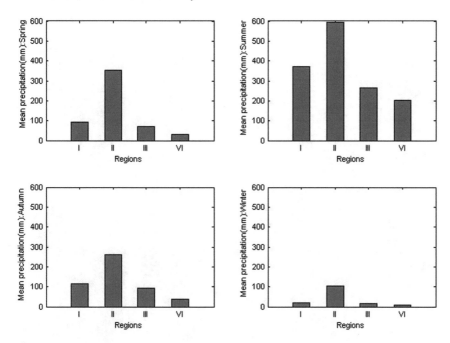

Fig. 6.2 Average seasonal precipitation from 1951 to 2010 in each subarea

Fujian Provinces and Nanling mountainous region, regions dominated by low hills and plains in Fujian, Guangdong and Guangxi Provinces, and the Hainan Peninsula, with the maximum average precipitation reaching 1348.08 mm. Less precipitation occurs in the Yunnan-Guizhou Plateau during spring, with an average precipitation of approximately 153 mm. The precipitation values in the Qinghai-Tibet Plateau and inland northwestern China are on average 67 mm and 31 mm, respectively, less than that in eastern China and consistent with that of Zhu and Sang (2018).

The summer monsoon is at its peak during summer, leading to generally increased precipitation in China. With the exception of slightly less precipitation in the inland areas of northwestern China, the average seasonal precipitation is greater than 100 mm in the southeastern half of China from the Greater Khingan Range to the lower reaches of the Brahmaputra River. The average seasonal precipitation in eastern China to the north of the Qinling Mountains and the Huaihe River is approximately 370 mm, which is higher than that in spring. The average summer precipitation in the area to the south of the Qinling Mountains and the Huaihe River is approximately 600 mm. Due to the influence of the southwest monsoon in summer, the summer precipitation in the Yunnan-Guizhou Plateau is significantly higher than that in spring, with the maximum average precipitation reaching 852 mm. Due to the influence of the southwest monsoon, the summer precipitation in the Qinghai-Tibet Plateau is significantly higher than that in spring, with the average summer precipitation in the south wing of the East Himalayas reaching approximately 1263 mm. With

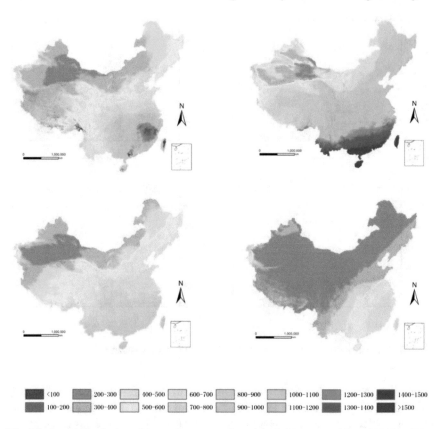

██ <100	██ 200–300	☐ 400–500	☐ 600–700	██ 800–900	☐ 1000–1100	██ 1200–1300	██ 1400–1500
██ 100–200	▒ 300–400	☐ 500–600	▒ 700–800	☐ 900–1000	☐ 1100–1200	██ 1300–1400	██ >1500

Fig. 6.3 Spatial distribution of the average seasonal precipitation between 1951 and 2010 in China (from the upper left to lower right: spring, summer, autumn, and winter)

an average amount of 205 mm, the summer precipitation in the Tianshan Mountains in northwestern China is significantly higher than that in spring.

In autumn, as the summer monsoon withdraws southward, the average precipitation in northwestern China is approximately 28 mm. Southwestern China and the southern coastal areas have more precipitation in autumn, with the average autumn precipitation in the subalpine areas of central and southern Yunnan Province and the mountainous areas in Xishuangbanna Dai Autonomous Prefecture reaching approximately 305 mm. The average autumn precipitation in eastern China to the south of the Qinling Mountains and the Huaihe River drops obviously compared with that in summer, with an average precipitation of approximately 263 mm. Compared with the summer precipitation, the autumn precipitation in eastern China to the north of the Qinling Mountains and the Huaihe River decreases by approximately 250 mm. The maximum precipitation in this subarea, 264 mm, is found at the southern foot of the Changbai Mountains.

As the winter monsoon is at its peak in winter, most areas of China experience the least precipitation in winter. The southeastern coastal areas have relatively more precipitation, with an average precipitation of approximately 164 mm near the Nanling mountains and the hilly regions to the south of the Yangtze River. In eastern China to the north of the Qinling Mountains and the Huaihe River, the maximum precipitation is still found around the Changbai Mountains, with an average precipitation of approximately 54 mm in winter. In western China, there is relatively more precipitation in the alpine valleys of south Tibet Autonomous Region and the mountainous regions in the Tianshan Mountains and Altai Mountains, and the average precipitation in the Tianshan Mountains is approximately 9 mm.

6.5 Results and Discussion

In view of the problems in interpolation of precipitation in China, a regression analysis of the seasonal precipitation data in different subareas is carried out with the consideration of the influences of geographical and topographic factors and local topographic factors around each grid point. Polynomial regression and stepwise regression are applied to identify the factors that significantly influence the precipitation in each subarea, and a total of 16 regression equations are established for each subarea during four seasons. Last, after the trends are removed, HASM is used to iteratively correct the residual. Since the key of HASM is to solve equations, the simulation accuracy of HASM is controlled by adjusting the iterative convergence accuracy of the equations while correcting the residual. In practical application, the interpolation results can be controlled based on the actual needs. In addition, when the HASM equations are iteratively solved, a buffer zone is introduced to ensure the simulation accuracy of HASM at the boundary. Moreover, the SSORCG method is used to accelerate the convergence of HASM based on the characteristics of the coefficient matrix of HASM equations, which not only shortens the computational time of HASM but also reduces the memory overhead.

The comparison of HASM with classical interpolation methods shows that HASM exhibits better performance in subareas with dense meteorological stations or sparse stations, indicating that HASM is applicable to simulating the seasonal precipitation in different areas of China. The average seasonal precipitation during 1951–2010 in different subareas is simulated by polynomial regression and HASM residual interpolation to reproduce the basic characteristics of the geographical distribution of and seasonal variations in precipitation in China. Based on the simulation results, the precipitation in China has strong regional and seasonal characteristics, and the precipitation is significantly influenced by local topographic factors. The regression analysis results show that the precipitation in China is mainly affected by the monsoon and that the influencing factors of precipitation in different regions vary with the seasons.

References

Bannayan M, Sanjani S. 2011. Weather conditions associated with irrigated crops in an arid and semi arid environment. Agricultural And Forest Meteorology, 151: 1589–1598.

Cai XM, Wang DB, Romain L. 2009. Impact of climate change on crop yield: A case study of Rainfed corn in central Illinois. Journal of Applied Meteorology and Climatology, 48: 1868–1881.

Daly C, Halbleib M, Smith JI, et al. 2008. Physiographically sensitive mapping of climatological temperature and precipitation across the conterminous United States. International Journal Of Climatology, 28: 2031–2064.

Deng ZX. 2006. Data analysis method and SAS System. Shanghai University of Finance and Economics Press.

Joly D, Brossard T, Cardot H, et al. 2011. Temperature interpolation based on local information: The example of France. International Journal Of Climatology, 31: 2141–2153.

Kleijen JP, van Beers WC. 2005. Robustness of kriging when interpolating in random simulation with heterogeneous variances: Some experiments. European Journal of Operational Research, 165: 826–834.

Li SK, Hou GL, Ouyang H, et al. 1988. Agroclimatic resources and agroclimatic regionalization in China. CRC Press.

Liu XA, Yu GR, Fan LS, et al. 2004. Study on spatialization of terrestrial eco-information in China(III):Temperature and precipitation. Journal of Natural Resources, 19: 818–825.

Marquinez J, Lastra J, Garcia P.2003. Estimation models for precipitation in mountainous regions: The use of GIS and multivariate analysis. Journal of Hydrologic Engineering, 270: 1–11.

Moulin L, Gaume E, Obled C. 2009. Uncertainties on mean areal precipitation: assessment andimpact on streamflow simulations.Hydrology and Earth Systrm Sciences,13: 99–114.

Schoonmakers SJ. 2003. The CAD Guidebook. New York: Marcel Dekker, Inc.

Shi WJ, Liu JY, Du ZP, et al. 2009. Surface modeling of soil PH. Geoderma, 150: 113–119.

Sinoweki W, Scheinost AC, Auerswald K. 1997. Regionalization of soil water retention curves in a highly variable soilscape: II. Comparison of regionalization procedures using a pedotransfer function. Geoderma, 78: 145–159.

Tian YZ, Yue TX, Zhu LF, et al. 2005. Modeling population density using land cover data. Ecological Modelling, 189: 72–88.

Wotling G, Bouvier C, Danloux J, et al. 2000. Regionalization of extreme precipitation distribution using the principal components of the topographical environment.Journal of hydrology, 233: 86–101.

Yan H, Nix HA, Hutchinson M F, et al. 2005. Spatial interpolation of monthly mean climate data for China. International Journal Of Climatology, 25: 1369–1379.

Yue TX, Du ZP, Liu JY. 2004. High precision surface modeling and error analysis. Progress in Natural Science, 14: 300–306.

Yue TX, Du ZP. 2006. Error comparison between high precision surface modeling and classical model. Progress in Natural Science, 16: 986-991.

Yue TX, Fan ZM, Liu JY. 2007. Scenarios of land cover in China. Global Planet Change, 55: 317–342.

Yue TX. 2011. Surface Modeling: High Accuracy and High Speed Methods. CRC Press.

Zhang S, Liao SB. 2011.Simulation and analysis of spatialization of mean annual air temperature based on BP neural network. Journal of Geoinformation Science, 2011, 13: 534–538.

Zhu YX, Sang YF. 2018. Spatial variability in the seasonal distribution of precipitation on the Tibetan Plateau. Progress in Geography, 37: 1533–1544.

Chapter 7
HASM of the Percentage of Sunshine in China

The percentage of sunshine is one of the important factors in the study of sunshine duration and solar radiation. Therefore, the simulation results of percentage of sunshine directly impact research and applications in the related fields. High-accuracy surface modelling (HASM) is a high-precision surface simulation method for ecological modelling developed in recent years. In this study, the existing HASM is modified and a more accurate surface modelling method with a complete theoretical basis is proposed, referred to as HASM.MOD. Using the Gaussian surface as an example, the simulation accuracy of HASM and HASM.MOD is compared. The experimental results show HASM.MOD has higher simulation accuracy than HASM. Then, based on the average monthly percentage of sunshine data collected at 752 meteorological stations in China from 1951 to 2010, HASM.MOD is used to simulate the distribution of average monthly percentage of sunshine in China, and the results are compared with those of the HASM, kriging, and inverse distance weighting (IDW) methods. The results showed that HASM.MOD has the highest accuracy for each month, and the percentage of sunshine data obtained by HASM.MOD can serve as reference geographic data for related studies.

7.1 Introduction

Percentage of sunshine is not only an important factor in studying sunshine duration and solar energy but also one of the main meteorological elements characterizing climate change. Sunshine duration is often calculated based on the percentage of sunshine, so the simulation accuracy of the percentage of sunshine directly determines the calculation accuracy of sunshine duration. Sunshine duration, as an important factor for the amount of solar energy received by a place, is a primary factor for the formation of local climate and an important factor for the distribution of plants. Sunshine duration affects the distribution of solar radiation and is closely related to agricultural production; therefore, it is one of the main climatic factors that determine

N. Zhao and T. Yue, *High Accuracy Surface Modeling Method: The Robustness*, https://doi.org/10.1007/978-981-16-4027-8_7

the ecological productivity of a region. Kimball et al. (1919) pointed out a possible relationship between percentage of sunshine and solar radiation. Angstorm (1924) presented a formula for calculating total solar radiation by percentage of sunshine. In a general survey of solar energy resources in various places, the total solar radiation on a horizontal plane is mostly estimated according to the percentage of sunshine data collected at meteorological stations (Huang et al. 2006; Hao et al. 2009; He and Xie 2010; Xin et al. 2011). Moreover, percentage of sunshine is the input parameter of ecological model such as the Lund-Potsdam-Jena (LPJ) model. Hence, the simulation accuracy of the percentage of sunshine greatly affects the simulation results of the LPJ model (Smith et al. 2001; Sitch et al. 2003).

The average percentage of sunshine reflects the influence of local climatic features on sunshine duration, including cloud coverage, visibility, and water vapor. Currently, there are few studies on the percentage of sunshine. Previous studies are mainly based on classical interpolation methods, such as the IDW method (Yuan et al. 2008; Pan 2012) and the kriging method (Wang et al. 2011). The accuracy of these interpolation methods is low, which often leads to large errors in the simulated percentage of sunshine, thereby affecting the calculation accuracy of sunshine duration and solar radiation.

HASM is a highly accurate spatial surface simulation method proposed in recent years for the simulation of geographic information systems (GISs) and ecosystems (Yue 2011). Numerical tests and practical applications have shown that HASM is more accurate than classical interpolation methods, such as the kriging, IDW and spline methods (Yue and Du 2006; Yue et al. 2007, 2011). Although HASM is a relatively accurate interpolation method, its simulation performance for percentage of sunshine has not been verified. In addition, the theoretical basis of HASM is incomplete, which affects the interpolation results. In this study, HASM is improved, and after the improvement of its theoretical basis, the modified HASM method is referred to as HASM.MOD. First, the accuracy of HASM.MOD is verified by simulating mathematical surfaces. Then, based on the average seasonal precipitation during 1951–2010, the accuracy of HASM.MOD is compared with those of HASM and the classical interpolation methods (i.e., kriging, IDW, and spline) to determine the most accurate method for simulating percentage of sunshine.

7.2 Data and Study Area

Based on the monthly observation data collected at 752 meteorological stations in China from 1951 to 2010, the average percentage of sunshine of each station is obtained. The longitude and latitude information of each station is also included. Meteorological stations are unevenly distributed in China, and stations are dense in eastern China and sparse and extremely uneven in western China (Fig. 7.1). The meteorological stations are constructed at different times. Before 1950, there were few meteorological stations to the west of 100°E in China, and large-scale construction of meteorological stations did not begin until the early 1950s. Through the analysis

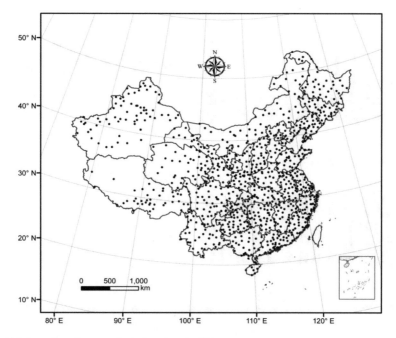

Fig. 7.1 Location of meteorological stations in China

of data from each station, the stations with missing data are removed, and a total of 740 stations are included in this study.

7.3   Simulation of Average Monthly Percentage of Sunshine

A total of 85% of the stations are randomly selected for simulation, and the remaining 15% are used for validation. The simulation is repeated 10 times, and the root mean square error (RMSE) values for different methods are shown in Table 7.1.

The simulation accuracy of HASM.MOD is higher than those of HASM and the classical interpolation methods for each month. The simulation accuracy of HASM

Table 7.1 Comparison of the accuracies of different methods

Month	1	2	3	4	5	6	7	8	9	10	11	12
HASM.MOD	5.09	4.10	3.76	3.06	2.80	3.28	3.44	3.44	3.35	3.61	4.16	5.25
HASM	5.28	4.45	4.16	3.52	3.16	3.44	3.95	4.39	3.49	3.76	4.29	5.36
IDW	5.21	4.37	4.7	3.48	3.24	3.46	4.15	4.67	3.55	3.79	4.25	5.32
Kriging	5.12	4.13	3.78	3.07	2.80	3.29	3.95	4.4	3.37	3.63	4.18	5.28
Spline	7.54	6.19	5.22	4.24	2.80	5.05	5.47	6.24	5.81	5.54	6.15	7.60

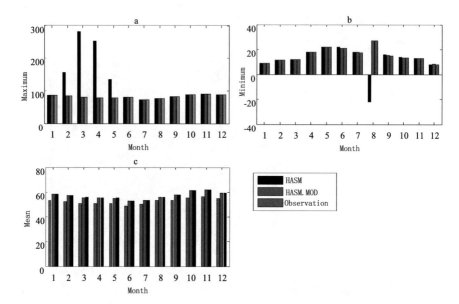

Fig. 7.2 Comparison between HASM.MOD and HASM. **a** Maximum. **b** Minimum. **c** Mean

is lower than that of the kriging method in months other than July and August. The spline method exhibits the lowest accuracy in all months except May. In addition, the simulation accuracy of each method in summer is better than that in winter.

The differences in the maximum, minimum and mean percentage of sunshine between each method and the measured values are shown in Fig. 7.2 for comparison of the simulation results of HASM.MOD and HASM. The maximum, minimum and mean values are obtained using the statistical analysis function of ArcGIS.

As shown in Fig. 7.2, the maximum values of HASM.MOD are very close to the measured values, and the maximum values of HASM in February, March, April, and May are largely different from the measured values. Figure 7.2b shows that the minimum value of HASM in August is negative, which is inconsistent with the measured value. Figure 7.2c shows the mean values of the two methods; the simulated values by HASM.MOD and HASM are larger than the measured values, and the values of HASM.MOD are closer to the actual values than those of HASM.

Figure 7.3 shows the distributions of percentage of sunshine in China in January, April, July, and October obtained by HASM.MOD.

As shown in Fig. 7.3, in January, the percentage of sunshine to the south of the Qinling Mountains and the Huaihe River is low; the percentage of sunshine is also relatively low in the Sichuan Basin, southern Chongqing Municipality, Guizhou Province, northern Guangxi Province and most areas of Hunan Province, and the South Tibet Autonomous Region has the highest percentage of sunshine. In April, except in most parts of Yunnan, the percentage of sunshine to the south of the Qinling Mountains and the Huaihe River is generally low, and the percentage of sunshine in Heilongjiang Province, southeastern Jilin Province, and the Tianshan Mountains is

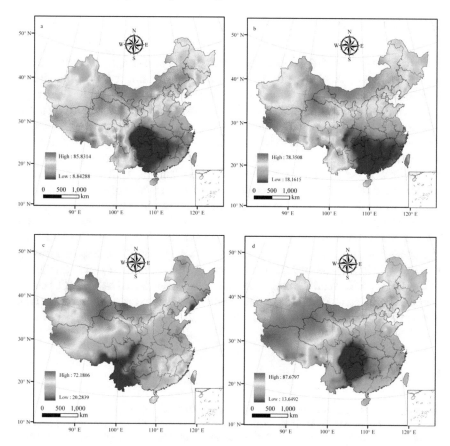

Fig. 7.3 Distributions of the percentage of sunshine obtained by HASM.MOD. **a** January. **b** April.
c July. **d** October

also low. Compared with that in January, the area with a low percentage of sunshine
tends to decrease in April. In July, the lowest percentage of sunshine is located in the
western part of the Yunnan-Guizhou Plateau and the southeastern Tibet Autonomous
Region, and the percentage of sunshine in southern Liaoning Province is also low.
Compared with April, the percentage of sunshine in southeastern China and southern
Xinjiang Uygur Autonomous Region increases in July, and the percentage of sunshine
in the Tibet Autonomous Region, Qinghai Province, Shaanxi Province and Hebei
Province decreases in July. In October, northwestern China and most parts of Inner
Mongolia Autonomous Region are the regions with a high percentage of sunshine,
while the lowest values are found in the Sichuan Basin, Guizhou Province and south
Chongqing Municipality. Compared with July, the lowest percentage of sunshine
shows a significant shift in October. In general, the percentage of sunshine in northwestern China is higher than that in the southeastern coastal areas, with the lowest
value near the Sichuan Basin.

7.4 Conclusion

With the average monthly percentage of sunshine as an example, the performance of HASM.MOD is verified, and the results show that HASM.MOD possesses higher simulation accuracy than HASM and the classical interpolation methods that are often used to simulate the percentage of sunshine. The current study presents an accurate interpolation method for future studies in related fields. Through a comparison of the differences in the maximum, minimum and mean values of HASM.MOD and HASM, the advantages of HASM.MOD over HASM are verified. Then, the distributions of percentage of sunshine in China in January, April, July and October are presented (Fig. 7.3). The results show that the percentage of sunshine in northwestern China is generally higher than that in the southeastern coastal areas and that the minimum value is found near the Sichuan Basin. Since the nonlinear equations satisfied by the mixed partial derivative are introduced in HASM.MOD, the computational time and memory overhead of HASM.MOD are approximately 6 times and 2 times those of HASM, respectively. In the next step, the parallelization of HASM.MOD can be implemented, and the parallelized HASM.MOD can be combined with high-resolution satellite remote sensing data to provide more accurate percentage of sunshine data.

References

Angstrom A. 1924. Solar and atmospheric radiation. Q J R Met Soc, (20): 6121–6126.

Hao CY, Xu CY, Wu SH. 2009. Spatialization of total solar radiation in mountain area of Southern Yunnan Based on DEM model and Climatological Calculation. Resource Science, 31(6): 1031–1039.

He QH, Xie Y. 2010. Climatological Calculation Method of solar radiation in China. Journal of natural resources, 25(2): 308–319.

Huang WH, Shuai XQ, Wang KJ. 2006. Study on GIS model of sunshine and radiation in mountain area considering terrain conditions. Chinese Journal of Agrometeorology, 27(2): 89–93.

Kimball HH. 1919. Variations in the total and luminous solar radiation with geographical position in the United States. Mon Wea Rev.47(11): 769–793.

Pan YD. 2012. The revision of the calculation model of sunshine time under the topography fluctuation. Resource Science, 32(8): 1493–1498.

Sitch S, Smith B, Prentice IC, et al. 2003. Evaluation of ecosystem dynamics, plant geography and terrestrial carbon cycling in the LPJ dynamic global vegetation model. Global Change Biology, 9: 161–185.

Smith B, Prentice IC, Sykes MT. 2001. Representation of vegetation dynamics in the modeling of terrestrial ecosystems: comparing two contrasting approaches within European climate space. Global Ecology and Biogeography, 10(6): 621–637.

Wang HQ, Yin JM, Zhan MJ, et al. 2011. A precise calculation model of sunshine hours considering terrain shading. Chinese Journal of Agrometeorology, 32(2): 273–278.

Xin Y, Zhao YZ, Mao WY, et al. 2011. Homogeneity test of total solar radiation data in Xinjiang and discussion of climatological estimation formula. Plateau Meteorology,30(4): 878–889.

Yuan SJ, Mu QL, Qiu XF, et al. 2008. The spatial distribution of sunshine time under the undulating terrain of Guizhou Plateau. Journal of Applied Meteorology, 19(2): 233–237.

Yue TX, Du ZP, Song DJ, et al. 2007. A new method of high accuracy surface modeling and its application to DEM construction. Geomorphology, 91: 161–172.

Yue TX, Du ZP. 2006. High accuracy surface modeling and comparative analysis of its errors. Prog Nat Sci 16: 986–991.

Yue TX, Fan ZM, Chen CF, et al. 2011. Surface modeling of global terrestrial ecosystems under three climate change scenarios. Ecological Modeling, 222: 2342–2361.

Yue TX. 2011. Surface modeling: high accuracy and high speed methods. CRC Press, New York.

Chapter 8
Simulation of Potential Evapotranspiration in the Heihe River Basin by HASM

Potential evapotranspiration (ET) is an important factor in the study of ET and regional water circulation, and the simulation accuracy of potential ET has a strong impact on the relevant studies and practical applications. In this study, the potential ET in the Heihe River Basin is simulated based on potential ET data collected four times a day at 12 meteorological stations in the Heihe River Basin from 2000 to 2009. First, according to the characteristics of the potential ET, the meteorological, geographical and topographic factors affecting the potential ET in the study area are analysed, and the background trend of potential ET in the study area is simulated by means of polynomial regression and stepwise regression to identify the optimal combination of factors influencing the potential ET. On this basis, high-accuracy surface modelling (HASM) is applied to correct the residual error after removing the trends. The simulation accuracy of HASM is verified by comparing with the potential ET data by the kriging, inverse distance weighting (IDW), and spline methods and the "3-km 6-h simulated meteorological forcing data in the Heihe River Basin from 1980 to 2010". The results showed that HASM yields the highest simulation accuracy, and the distribution results are reasonable, indicating a promising application prospect of HASM in potential ET simulation. The potential ET data obtained by HASM can serve as reference geographic data in related studies.

8.1 Introduction

As a key factor in regional water circulation and an important part of surface heat balance, ET can be divided into three categories, i.e., pan ET, potential ET, and actual ET. Potential ET, as a measure of the atmospheric ET capacity, is an important index for estimating the drought level and crop water consumption (Liu et al. 2009; Zhu et al. 2011; Hao et al. 2013). Accurate simulation of the spatial distribution of potential ET is of great significance for hydrological research, water resource

© The Author(s), under exclusive license to Springer Nature Singapore Pte Ltd. 2021   147
N. Zhao and T. Yue, *High Accuracy Surface Modeling Method: The Robustness*,
https://doi.org/10.1007/978-981-16-4027-8_8

assessment, regional water circulation changes, and climate regionalization under the background of contemporary climate change.

Scholars in China and abroad have conducted a large number of studies. The Penman–Monteith model recommended by the Food and Agriculture Organization (FAO) of the United Nations is a commonly used method for estimating potential ET (Allen et al. 1998). Xu et al. (2014) used the meso-scale Weather Research and Forecasting (WRF) model to simulate the water vapor budget of the Qinghai-Tibet Plateau. Wu et al. (2008) used ETWatch, an operational system for monitoring regional ET with remote sensing, to simulate ET through remote sensing. With the development of geographic information system (GIS) technology in recent years, the spatial interpolation methods used in the corresponding software, such as the IDW (Wang et al 2011), kriging (Yuan et al. 2008) and spline (Wu et al. 2004) methods, have provided new approaches for the spatial simulation of meteorological elements. The kriging method requires variables to meet second-order stationary conditions, which is hard to realize in practical applications, so the simulation accuracy is negatively affected. IDW is sensitive to distance and requires the sampling points to be uniformly distributed and clustered over the whole region. The spline method is suitable only for simulating smooth surfaces. Thus, the classical interpolation methods are often limited by the application conditions, so the simulation accuracy of these methods is affected (Qian et al. 2010; Du 2013).

Proposed in the 1990s, HASM has been developed into a mature and highly accurate surface modelling method (Yue 2011). Based on the principal theorem of surface theory in differential geometry, HASM solves problems caused by errors and multiple scales in surface modelling. HASM has been successfully applied to the fields of population distribution, soil pH value, and climate, and its simulation accuracy is higher than those of classical interpolation methods (Yue et al. 2004; Zhao et al. 2013; Zhao et al. 2015). In this study, HASM is used to simulate the potential ET in the Heihe River Basin, and the results are compared with those obtained by other methods to evaluate the simulation accuracy of HASM.

8.2   Study Area and Data Source

As the second largest inland river in China, the Heihe River originates from the Qilian Mountains in the northeast part of the Qinghai-Tibet Plateau and flows through Gansu Province, Inner Mongolia Autonomous Region and Qinghai Province. Starting from the Shiyang River Basin in the east, reaching the Shule River Basin in the west, facing the Qilian Mountains in the south, and stretching to the border between China and Mongolia in the north, the Heihe River Basin is located at $37°50'–42°40'$ N and $97°30'–102°$ E (Lu et al. 2012). The main stream is 821 km long, and the basin covers an area of 12.8×10^4 km^2. The elevation is high in the southwestern Heihe River Basin and low in the northeast. The basin is divided into the Gobi, plain, and mountainous areas from north to south. The climate is dry with scarce and concentrated precipitation in the basin (Liao et al. 2015).

The potential ET data collected four times a day at 12 meteorological stations in the Heihe River Basin from 2000 to 2009 are used in this study. Based on the remote sensing topographic data of the Heihe River Basin, the potential ET in the Heihe River Basin from 2000 to 2009 is simulated using HASM. The 12 meteorological stations are located in Sunan County, Jiuquan City, Mazongshan Town, Yumen Town, Dingxin Town, Jinta County, Gaotai County, Linze County, Zhangye City, Minle County, Shandan County and Yongchang County. The stations are unevenly distributed, and mostly concentrated in the southeast and middle of the Basin (Fig. 8.1).

In this study, the "3-km 6-h simulated meteorological forcing data in the Heihe River Basin from 1980 to 2010" are used as the control data. The data are extracted from the Heihe River Project Data Management Center (http://www.heihedata. org/). The data are based on the Regional Integrated Environmental Model System (RIEMS 2.0) developed by the Key Laboratory of Regional Climate-Environment for Temperate East Asia (RCE-TEA), Chinese Academy of Sciences. The important parameters in RIEMS 2.0 are recalibrated by the observation and remote sensing data of the Heihe River Basin. The vegetation data are based on the 2000 land use

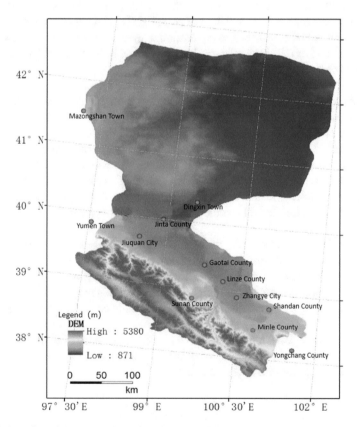

Fig. 8.1 Location of the meteorological stations in the Heihe River Basin

data and 30 s digital elevation model (DEM) data of the Heihe River Basin in the Heihe River Basin data inventory. In this way, a regional climate model suitable for the study of eco-hydrological processes in the Heihe River Basin is established. The data are unformatted data stored in grads format. The data are converted into ASCII data and loaded into ArcGIS for extraction. Moreover, the accuracy of HASM is evaluated by comparing the simulation results with those of classical interpolation methods (e.g., kriging, IDW and spline).

8.3 Methods

8.3.1 Polynomial Regression and Stepwise Regression

The distribution of a variable during a spatial process is the result of the comprehensive influences of multiple factors. During this spatial process, the changes in the variable come from 2 parts, i.e., fixed global trends and random short-range variations. To realize the spatial simulation of potential ET within a given period of time, equations should be established to represent the regression relationship between the potential ET and its influencing factors based on an analysis of the influencing factors in the study area.

In many studies, the influences of meteorological factors, such as the average temperature, average relative humidity, precipitation, average wind speed and sunshine duration, on the potential ET are considered (Liu et al. 2009; Zhu et al. 2011; Hao et al. 2013). Wang et al. (2010) reported that the trend and amplitude of the potential ET and their main influencing factors varied from one region to another because of different topographic features and geographical locations. The influencing factors of the potential ET investigated in this study include the humidity (%), wind speed (m/s), sunshine duration (h), temperature (°C), precipitation (mm), longitude (m), latitude (m), elevation (m), slope (°), aspect (°) and topographic relief (m).

The simulation of potential ET based on its influencing factors varies with climate, topography, and time. Therefore, polynomial regression is used to fit the relationship between the potential ET and climatic and topographic factors, and then stepwise regression is applied to select the optimal subset of influencing factors (Deng 2006). All the influencing factors in the subset are significantly correlated with the potential ET.

Geographic weighted regression (GWR) is used to simulate the mean coefficient of variation (CV) of the regression coefficients to analyse the spatial stationarity of the regression relationship of the potential ET.

$$\overline{CV} = \left(\sum_{i=1}^{n} \frac{S_i}{\overline{X_i}} \times 100\% \right) / n \tag{8.1}$$

where n denotes the number of regression coefficients, i represents the i-th regression coefficient, S_i is the standard deviation of the i-th regression coefficient, and $\overline{X_i}$ refers to the mean attribute value. If $\overline{CV} < 10\%$, the spatial variability is weak; if $10\% \leq \overline{CV} < 100\%$, the spatial variability is moderate; and if $\overline{CV} > 100\%$, the spatial variability is strong. When $\overline{CV} < 100\%$, the spatial variability of the potential ET is weak, and the spatial distribution is stable. Hence, the ordinary least squares (OLS) can be used to obtain the trend surface of ET in the study area; when $\overline{CV} \geq 100\%$, the spatial variability of the potential ET is strong, and the spatial distribution is unstable. Therefore, GWR can be applied to obtain the trend surface of the ET in the study area. The trend surface serves as the driving field for follow-up studies. This step is carried out in ArcGIS software.

Table 8.1 shows the regression equations and CVs of average monthly ET data in the Heihe River Basin from 2000 to 2009, where X represents the longitude, Y denotes the latitude, DEM represents the elevation, Slope stands for the slope, Aspect is the aspect, Wind is the wind speed, Tem denotes the temperature, Pre represents the precipitation, and Humidity represents the humidity. As shown in Table 8.1, as the climatic factors change, the main influencing factors of potential ET are different in different months.

8.3.2 HASM

Based on the principle of differential geometry and the optimal control theory, Yue et al. (2004) established a HASM method with global approximate data (including remote sensing data and coarse resolution simulation data by global models) as the driving field and local high-precision data (including monitoring network data and survey sampling data) as optimal control conditions.

The principal theorem of surface theory is as follows. Let the first and second fundamental coefficients E, F, G, L, M and N be symmetric, E, F and G be positive definite, and E, F, G, L, M and N satisfy the Gaussian equations. Then, $z = f(x, y)$ is the one unique solution to the total differential equations under the initial condition $f(x, y) = f(x_0, y_0)(x = x_0, y = y_0)$.

The Gaussian equations are expressed as follows:

$$\begin{cases} f_{xx} = \Gamma f_{11}^1 f_x + \Gamma f_{11}^2 f_y + \frac{L}{\sqrt{E+G-1}} \\ f_{yy} = \Gamma f_{22}^1 f_x + \Gamma f_{22}^2 f_y + \frac{N}{\sqrt{E+G-1}} \\ f_{xy} = \Gamma f_{12}^1 f_x + \Gamma f_{12}^2 f_y + \frac{M}{\sqrt{E+G-1}} \end{cases} \tag{8.2}$$

where $E = 1 + f_x^2$, $F = f_x \cdot f_y$, $G = 1 + f_y^2$,
$L = \frac{f_{xx}}{\sqrt{1+f_x^2+f_y^2}}$, $M = \frac{f_{xy}}{\sqrt{1+f_x^2+f_y^2}}$, $N = \frac{f_{yy}}{\sqrt{1+f_x^2+f_y^2}}$,
$\Gamma_{11}^1 = \frac{1}{2}(GE_x - 2FF_x + FE_y)(EG - F^2)^{-1}$, $\Gamma_{11}^2 = \frac{1}{2}(2EF_x - EF_y - FE_x)(EG - F^2)^{-1}$,

Table 8.1 Average monthly ET data in the Heihe River Basin in 2000–2009

Month	CV (%)	Spatial stationarity	Trend surface	Regression relationship	R^2	Adjusted R^2
1	0.0601	Stationary	OLS	Eva = 85.891893 + 0.000038 × X-0.000011 × Y + 1.42571 × Slope + 3.74786 × Wind-2.191041 × Tem	0.7943	0.5476
2	0.8197	Stationary	OLS	Eva = 442.704724 + 0.000054 × X + 0.000124 × Y + 0.020376 × DEM-5.210942 × Slope	0.6439	0.5188
3	0.0452	Stationary	OLS	Eva = 809.960102 + 0.000007 × X + 0.000216 × Y + 0.070939 × DEM-23.591797 × Slope + 0.032568 × Aspect-10.638409 × Pre	0.9428	0.8741
4	0.2556	Stationary	OLS	Eva = 2021.251854 + 0.0001 × X + 0.000529 × Y + 0.119819 × D EM-1.484043 × Focalst-2.130829 × Wind-7.674418 × Pre	0.9472	0.8839
5	1.3382	Stationary	OLS	Eva = 1965.742178 + 0.000538 × X + 0.000647 × Y + 0.052113 × DEM + 5.869211 × Slope-37.088237 × Wind-2.967979 × Humidity	0.8733	0.7678
6	0.0711	Stationary	OLS	Eva = 3495.605701 + 0.00055 × X + 0.000982 × Y + 0.070291 × DEM-4.273307 × Slope-47.9458 × Wind	0.8390	0.7049
7	0.0702	Stationary	OLS	Eva = 2205.70824 + 0.000507 × X + 0.000702 × Y + 0.043407 × DEM + 5.458829 × Slope-2.241815 × Humidity- 41.872033 × Wind	0.9491	0.8879
8	0.1780	Stationary	OLS	Eva = 1531.148667 + 0.000398 × X + 0.000512 × Y + 0.037061 × DEM + 3.266468 × Slope-2.055309 × Humidity-35.054072 × Wind	0.9092	0.8003
9	0.0313	Stationary	OLS	Eva = 813.651958 + 0.000173 × X + 0.000285 × Y + 8.544999 × Slope + 1.791287 × Humidity-23.872159 × Wind + 0.682247 × Pre	0.9071	0.7958
10	19.2299	Stationary	OLS	Eva = 2895.536798 + 0.000473 × X + 0.000763 × Y + 0.154767 × DEM-1.353498 × Focalst + 1.257364 × Humidity-12.786523 × Wind-10.088294 × Pre	0.7445	0.5579
11	0.0289	Stationary	OLS	Eva = 6.559405 + 0.000082 × X + 0.00004 × Y + 0.041354 × Aspect-0.827806 × Humidity + 7.86809 × Wind	0.8307	0.6897
12	0.0246	Stationary	OLS	Eva = -61.675279 + 0.00007 × X + 0.00003 × Y + 2.144228 × Slope + 0.053166 × Humidity-3.872257 × Tem	0.8623	0.6971

Table 8.2 HASM error analysis (unit: mm)

Code	Mazongshan town	Yumen town	Dingxin town	Jinta county	Jiuquan city	Gaotai county	Linze county	Sunann county	Zhangye city	Minle county	Shandan county	Yongchang county
RMSE	55.18	47.57	44.74	18.36	23.80	19.54	69.50	52.77	27.01	47.40	23.09	49.95
MAE	43.94	30.17	36.11	15.58	17.79	17.17	47.72	35.95	17.95	42.64	17.43	39.92
MRE	0.25	0.19	0.20	0.12	0.15	0.17	0.23	0.21	0.13	0.34	0.13	0.31

Table 8.3 Error analysis of the meteorological forcing data (unit: mm)

Code	Mazongshan town	Yumen town	Dingxin town	Jinta county	Jiuquan city	Gaotai county	Linze county	Sunann county	Zhangye city	Minle county	Shandan county	Yongchang county
RMSE	148.82	95.39	153.42	95.21	91.8	88.67	444.77	124.31	195.44	208.60	157.83	189.31
MAE	130.10	73.94	115.09	80.89	63.5	64.27	329.89	75.06	115.94	145.05	105.05	130.08
MRE	0.72	0.55	0.79	0.63	0.61	0.62	1.97	0.55	1.01	0.89	0.70	1.01

Table 8.4 Error analysis of the kriging method (unit: mm)

Code	Mazongshan town	Yumen town	Dingxin town	Jinta county	Jiuquan city	Gaotai county	Linze county	Sunann county	Zhangye city	Minle county	Shandan county	Yongchang county
RMSE	64.12	30.32	47.11	26.69	35.34	19.08	66.72	20.35	29.84	47.60	26.97	56.71
MAE	47.68	24.24	37.75	19.30	28.17	17.98	47.64	17.21	21.74	41.67	22.24	41.64
MRE	0.29	0.11	0.20	0.14	0.16	0.16	0.24	0.12	0.12	0.28	0.13	0.27

Table 8.5 Error analysis of the IDW method (unit: mm)

Code	Mazongshan town	Yumen town	Dingxin town	Jinta county	Jiuquan city	Gaotai county	Linze county	Sunann county	Zhangye city	Minle county	Shandan county	Yongchang county
RMSE	55.60	16.26	52.14	31.68	29.40	34.45	70.08	30.42	48.79	46.78	24.71	41.46
MAE	44.57	12.81	40.25	27.32	24.02	27.97	48.75	20.16	32.94	40.91	19.61	38.54
MRE	0.27	0.08	0.22	0.16	0.14	0.17	0.24	0.13	0.16	0.27	0.12	0.24

Table 8.6 Error analysis of the spline method (unit: mm)

Code	Mazongshan town	Yumen town	Dingxin town	Jinta county	Jiuquan city	Gaotai county	Linze county	Sunann county	Zhangye city	Minle county	Shandan county	Yongchang county
RMSE	187.85	56.81	243.1	30.57	269.9	121.2	84.3	86.38	90.93	122.5	96.59	244.6
MAE	139.95	43.95	86.17	24.53	96.73	86.29	58.65	50.93	61.96	79.15	68.11	155.5
MRE	0.86	0.23	0.52	0.14	0.57	0.43	0.29	0.35	0.31	0.39	0.35	0.76

$\Gamma_{22}^1 = \frac{1}{2}(2GF_y - GG_x - FG_y)(EG - F^2)^{-1}, \Gamma_{22}^2 = \frac{1}{2}(2EG_y - 2FF_y + FG_x)(EG - F^2)^{-1}$, and.

$\Gamma_{12}^1 = \frac{1}{2}(GE_y - FG_x)(EG - F^2)^{-1}, \Gamma_{12}^2 = \frac{1}{2}(EG_x - FE_y)(EG - F^2)^{-1}$.

With the replacement of the partial derivative in Equation set (1) with the discrete difference, the finite difference form of the Gaussian equations is expressed as,

$$
\begin{cases}
\frac{f_{i+1,j} - 2f_{i,j} + f_{i-1,j}}{h^2} = \left(\Gamma_{11}^1\right)_{i,j} \frac{f_{i+1,j} - f_{i-1,j}}{2h} + \left(\Gamma_{11}^2\right)_{i,j} \frac{f_{i,j+1} - f_{i,j-1}}{2h} + \frac{L_{i,j}}{\sqrt{E_{i,j} + G_{i,j} - 1}} \\[2mm]
\frac{f_{i,j+1} - 2f_{i,j} + f_{i,j-1}}{h^2} = \left(\Gamma_{22}^1\right)_{i,j} \frac{f_{i+1,j} - f_{i-1,j}}{2h} + \left(\Gamma_{22}^2\right)_{i,j} \frac{f_{i,j+1} - f_{i,j-1}}{2h} + \frac{N_{i,j}}{\sqrt{E_{i,j} + G_{i,j} - 1}} \\[2mm]
\frac{f_{i+1,j+1} - f_{i+1,j} - f_{i,j+1} + 2f_{i,j} - f_{i-1,j} - f_{i,j-1} + f_{i-1,j-1}}{2h^2} = \\
\qquad \left(\Gamma_{12}^1\right)_{i,j} \frac{f_{i+1,j} - f_{i-1,j}}{2h} + \left(\Gamma_{12}^2\right)_{i,j} \frac{f_{i,j+1} - f_{i,j-1}}{2h} + \frac{M_{i,j}}{\sqrt{E_{i,j} + G_{i,j} - 1}}
\end{cases} \quad (8.3)
$$

A_1, A_2 and A_3 denote the coefficient matrices formed by the terms at the left ends of Equation set (2), and d, p and q represent the constant vectors of the terms at the right ends of the equations. According to the information of the sampling points, Equation set (2) is transformed into the following least squares problem:

$$
\begin{cases}
min \left\| \begin{bmatrix} A_1 \\ A_2 \\ A_3 \end{bmatrix} x - \begin{bmatrix} d \\ q \\ p \end{bmatrix} \right\| \\
s.t. Sx = k
\end{cases} \quad (8.4)
$$

where S and k are the sampling matrix and sampling vector, respectively; if $\widetilde{f_{i,j}}$ is the value of $z = f(x, y)$ at the p-th sampling point (x_i, y_j), then $S_{p,(i-1) \times J + j} = 1$, and $k_p = \widetilde{f_{i,j}}$.

HASM is finally transformed into a constrained least square problem controlled by ground sampling information, and the target is to minimize the overall simulation error so that the simulated value at the sampling point equals the sampling value. Making full use of sampling information is also an effective means of ensuring that the iteration approaches the optimal simulation result (Golub and Loan 2009).

To use HASM for simulation, the driving field and precision control points should be used as the input data (Zhao and Yue 2016) The driving field is the trend surface of the study area fitted by polynomial regression, and the precision control points refer to the high-accuracy data obtained by measurement (Zhao et al. 2013). The simulation process is divided into the following steps:

(1)   The original data collected at the meteorological stations from 2000 to 2009 are used to calculate the average monthly potential ET data in the study area. With the values from the stations as the precision control points, the first and second fundamental coefficients, the Christoffel symbols of the second kind, and constant vectors of d, p, q, and k are calculated according to the trend surface data of the potential ET.

(2) Coefficient matrices A_1, A_2, and A_3 are generated based on Eqs. (8.2) and (8.3). Coefficient matrix S of the constraint equations of the precision control points is generated according to the potential ET data collected by the stations, and the equations are then solved. Once the required accuracy is reached, the output result is a simulated surface combining the potential ET trend surface data and the potential ET data observed by the station (Zhao et al. 2014).

(3) Steps (1)–(2) are repeated to obtain the simulated diagram for each month.

8.4 Results

8.4.1 Accuracy

8.4.1.1 Error Verification

The HASM simulated values, meteorological forcing data and the simulated values by other interpolation methods are extracted, and the mean absolute error (MAE), root mean square error (RMSE) and mean relative error (MRE) are used as accuracy verification indices to compare the performance of these methods. The MAE, RMSE, and MRE are calculated as follows:

$$RMSE = \sqrt{\frac{\sum_{i=1}^{N}\left(f_{ij} - \widetilde{f}_{ij}\right)^2}{N}} \tag{8.5}$$

$$MAE = \frac{1}{N}\sum_{i=1}^{N}\left|f_{ij} - \widetilde{f}_{ij}\right| \tag{8.6}$$

$$MRE = \frac{1}{N}\sum_{i=1}^{N}\frac{\left|f_{ij} - \widetilde{f}_{ij}\right|}{f_{ij}} \tag{8.7}$$

f_{ij} denotes the observation value at the meteorological stations, and $\widetilde{f}_{ij}$ is the simulated value.

The cross-verification method is used, and the experiment is repeated 10 times. The mean values are calculated, and the errors of different methods are obtained. The results are shown in Tables 8.2, 8.3, 8.4, 8.5 and 8.6.

As shown in Tables 8.2, 8.3, 8.4, 8.5 and 8.6, the interpolation results of HASM are more accurate than those of meteorological forcing data and the spline method in the study area. The accuracy of HASM is higher than those of kriging and IDW for all stations except those in Sunan County, Yumen Town and Yongchang County. These three stations are located at the northern boundary of the Qilian Mountains and the upper reaches of the Heihe River, where the elevation difference is large, and the topographic relief is complicated. In addition to the geographical and topographic factors and meteorological factors, such as temperature and humidity, the potential ET in this area is influenced by the high-altitude mountain climate. Therefore, the simulation accuracy of HASM is not ideal. Since the simulation surface is smooth, the RMSE of the kriging method is slightly smaller than those of IDW and spline. For the spline method, the maximum simulated value is greater than the observation value, so it had the largest RMSE, i.e., the lowest accuracy, of all the methods.

8.4.1.2   Comparison of Average Monthly ET Data of Four Stations

The potential ET values of HASM and other methods at the Mazongshan station in the lower reaches of the Heihe River Basin and the Yumen Town station, Jiuquan station and Zhangye station in the upper and middle reaches are extracted and compared with the observation values at each station. The average monthly ET data of each station are shown in Figs. 8.2, 8.3, 8.4 and 8.5

Figure 8.2 shows a comparison of the average monthly ET data of Mazongshan station. Mazongshan station is the only meteorological station in the lower reaches of the Heihe River Basin. The simulation results of HASM are quite close to the average observation values at the station in term of both trend and magnitude. The interpolation results of kriging and IDW are smaller than the observation values from May to October. The interpolation results of spline are greater than the observation values to various degrees. The meteorological forcing data are lower than the observation values to various degrees. Figure 8.3 shows a comparison of the average monthly ET data of Yumen Town station. Yumen Town station is a meteorological station

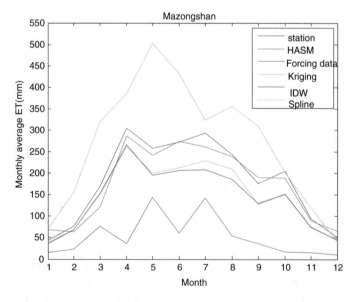

Fig. 8.2   Comparison of average monthly ET data of Mazongshan station

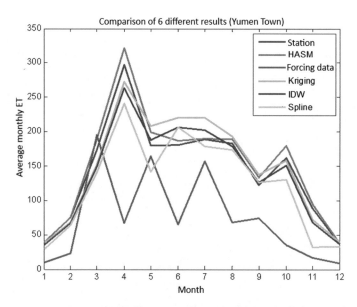

Fig. 8.3 Comparison of average monthly ET data of Yumen Town station

in the west of the Heihe River Basin. The results of HASM and IDW are relatively close to the average observation values of the station in terms of both trend and magnitude. The interpolation results of kriging are larger than the observation values from May to July. The interpolation results of spline are slightly smaller than the observation values as a whole, and the meteorological forcing data are lower than the observation values to various degrees. Figure 8.4 shows a comparison of the average monthly ET data of Jiuquan station. Jiuquan station is a meteorological station in the central area in the middle reaches of the Heihe River Basin. The simulation results of HASM are quite close to the average observation values. The interpolation results of kriging, IDW and spline are slightly larger than the observation values from May to September, but the overall trend is consistent with that of observation values. The meteorological forcing data greatly differ from the observation values from May to August. Figure 8.5 shows a comparison of the average monthly ET data of Zhangye station. Zhangye station is a meteorological station in the southeast of the basin. From June to September, the simulation results of HASM are slightly greater than the average observation values but smaller than those of kriging, IDW and spline, and the meteorological forcing data are mostly higher than the observation values. The simulated values of HASM are the closest to the observation values. In addition, since the potential ET data are relatively large in summer, there are certain deviations for different methods.

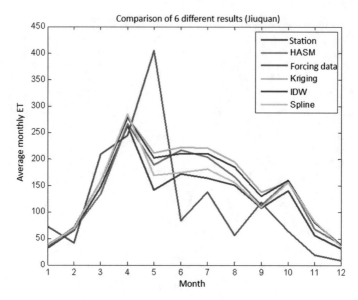

Fig. 8.4   Comparison of average monthly ET data of Jiuquan station

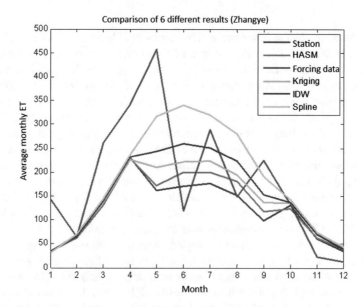

Fig. 8.5   Comparison of average monthly ET data of Zhangye station

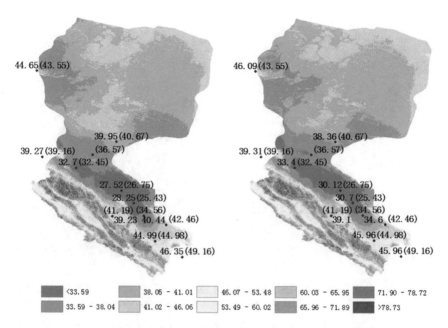

Fig. 8.6 Simulation results of potential ET in the Heihe River Basin in January

8.4.2 Comparison of the Potential ET in the Heihe River Basin Simulated by HASM and Kriging

The potential ET data in the Heihe River Basin obtained by HASM and kriging in January and July are compared in ArcGIS. The simulation results are shown in Figs. 8.6 and 8.7; the left shows the HASM results, and the right shows the kriging results. The values in parentheses refer to the observation values (unit: mm).

As shown in Figs. 8.6 and 8.7, although the results of HASM and kriging in January and July differ from the observation results in some areas in the middle and lower reaches of the Heihe River Basin, the overall distribution is roughly the same for the two methods, that is, high ET values in the south and low values in the north with the Qilian Mountains as the boundary. In Fig. 8.6, the potential ET of HASM is 43.6 mm, and that of kriging is 39.3 mm. At the border between Gansu Province and the Inner Mongolia Autonomous Region in the middle and lower reaches of the Heihe River Basin, the simulated value of kriging is slightly lower than that of HASM. As shown in Fig. 8.7, the potential ET obtained by HASM is 225.7 mm, and that of kriging is 256.1 mm. In the Inner Mongolia Autonomous Region in the middle and lower reaches of the Heihe River Basin, the simulated values of kriging are larger than those of HASM. Particularly, the simulated values of kriging in areas surrounding Mazongshan Town are larger than those of HASM on a year-on-year basis.

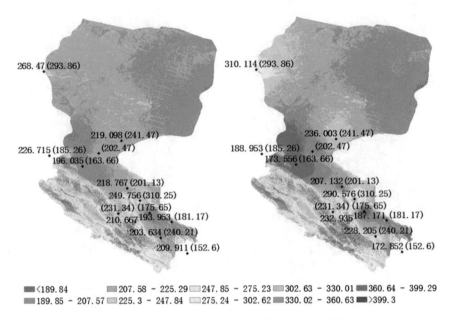

■<189.84   ■207.58 - 225.29▢247.85 - 275.23 ▢302.63 - 330.01■360.64 - 399.29
■189.85 - 207.57 ▢225.3 - 247.84   ▢275.24 - 302.62■330.02 - 360.63■>399.3

Fig. 8.7 Simulation results of potential ET in the Heihe River Basin in July

The monthly potential ET results obtained by the six methods are compared (Table 8.7). The second row shows the observation values at the meteorological stations, and rows 3–7 are the results of HASM, meteorological forcing data, kriging, IDW and spline, respectively.

As shown in Table 8.7, the potential ET results obtained by HASM are the closest to the observation values, followed by those of kriging and IDW. The simulated values of kriging are generally close to the observation values in April and May, and the simulated values of IDW are generally close to the observation values in March and June. The annual potential ET measured at the station is 1729.9 mm. The annual potential ET values simulated by HASM, meteorological forcing data, kriging, IDW, and spline are 1847.6, 1282.5, 1868.8, 1899.5, and 2589.4 mm, respectively.

Based on the above results, the simulation results of HASM are close to the observation values with small error indices and a high accuracy.

8.5   Results and Discussion

In this study, the potential ET in the Heihe River Basin from 2000 to 2009 is simulated. Using the trend surface obtained by polynomial regression analysis as the driving field, the HASM method is used correct the residual. The results are as follows:

The potential ET in Heihe River Basin is closely related to meteorological elements and geographical and topographic factors, and in general, there is a

Table 8.7 Comparison of the monthly potential ET results from the six methods (unit: mm)

Month	1	2	3	4	5	6	7	8	9	10	11	12
Station	38.1	68.2	147.8	252.4	205.3	217.5	214.9	190.6	132.6	149.5	73.6	39.4
HASM	43.6	68.3	169.9	262.5	223.2	208.9	225.7	218.6	153	154.3	79.3	40.3
Meteorological forcing data	36.6	22.4	93.5	79.2	244.9	125.8	323.6	146	126.3	54.8	19.8	9.59
Kriging	45.6	67.3	175.9	247.4	220.6	202.8	229.1	227.7	161.1	157.7	84	49.6
IDW	44.8	70.4	167.9	274.3	225.8	215.2	236.4	219.5	155.9	159.9	79.5	49.9
Spline	44.1	83.4	153.8	362.2	394	319.5	368.3	249.3	271.5	208.2	105.7	29.4

decreasing trend from south to north. The potential ET in the Qilian Mountains in the upper reaches of the basin is significantly larger than that in the middle and lower reaches of the basin. Moreover, the average monthly potential ET in the basin is related to seasonal changes. The lowest potential ET, 152.2 mm, is found in winter (December-February), followed by 386.6 mm in autumn (September–November), and the potential ET values in spring (March–May) and summer (June–August) are relatively high, 655.6 and 653.2 mm, respectively. The minimum average monthly potential ET, 40.3 mm, is found in December, and the maximum value of 225.7 mm is from July. The average annual potential ET in the basin reaches 1847.6 mm. Based on the above results, HASM has relatively high simulation accuracy and is suitable for the simulation of potential ET in the Heihe River Basin.

Through the simulation of potential ET, the performance of HASM in simulating climate change is verified. Compared with traditional interpolation methods, HASM yields better simulation performance. However, because of the complex computational process, the time consumption of HASM is high. In the simulation of meteorological elements, there is often not enough sampling data to reflect the spatial distribution of the meteorological elements. Hence, integrating HASM with factors influencing the spatial distribution of simulated meteorological elements and achieving parallel computing can be one of the main research directions in the next step.

References

Allen RG, Pereira LS, Raes D, et al. 1998. Crop Evapotranspiration-Guidelines for Computing Crop Water Requirements. FAO Irrigation and drainage paper, 56:100–109.

Deng ZX. 2006. Data analysis and SAS system. Shanghai: The MIT Press.

Du CZ. 2013. Comparative study of traditional temperature interpolation method based on GIS—Taking multi-years average temperature in Shandong Province as an example. Journal of Anhui Agricultural Sciences, 41(33):12939–12941.

Golub GH, Van Loan CF. 2009. Matrix computations. Beijing: Posts & Telecom Press.

Hao ZC, Yang RR, Chen XM, et al. 2013. Spatio-temporal patterns of the potential evapotranspiration in the Yangtze River catchment for the period of 1960–2011. Journal of Glaciology and Geocryology, 35 (2):408–418.

Liao J, Wang T, Xue X. 2015. Lake Evapotranspiration in the Ejin Basin since transfering water from the Heihe River. Journal of Desert Research, 35(1):328–332.

Liu M, Shen YJ, Zeng Y, et al. 2009. Changing trend of pan evapotranspiration and its cause over the past 50 years in China. Acta Geographica Sinica,64(3):259–269.

Lu GH, Xu D, He H. 2012. Characteristics of water vapor transportation and budget over the Heihe drainage basin. Journal of Natural Resources, 27(3):510–521.

Qian YL, Lv H Q, Zhang YH. 2010. Application and assessment of spatial interpolation method on daily meteorological elements based on ANUSPLIN software. Journal of Meteorology and Environment, 26(2):7–15.

Wang HQ, Yin JM, Zhan MJ, et al. 2011. Refining prediction model of sunshine hours considering terrain Masking. Chinese Journal of Agro-meteorology, 32(2):273–278.

Wang S , Zhang CJ, Han YX. 2010. Trend of potential evapotranspiration and pan evapotranspiration and their main impact factors in different climate regions of Gansu Province. Journal of Desert Research, 30(3):675–680.

Wu BF, Xiong J, Yan NN, et al. 2008. ETWatch for monitoring regional evapotranspiration with remote sensing. Advances in Water Science, 19(5):671–678.

Wu Y, Sun HY, Ma XZ. 2004. Semiparametric regression with cubic splin. Geomatics and Information Science of Wuhan University, 29(5):398–401.

Xu JY, Wang HJ, Li HY. 2014. Preliminary simulation analysis of water vapor budget of Qinghai-Xizang Plateau in summer. Plateau Meteorology, 33(5): 1173–1181.

Yuan SJ, Mao QL, Qiu XF, et al. 2008. The spatial and temporal distribution of insolation duration over rugged terrains in the Guizhou Plateau. Journal of Applied Meteorological Science, 19(2):233–236.

Yue TX, Du ZP, Liu JY. 2004. High precision surface modeling and error analysis. Progress in Natural Science, 14(3): 300–306.

Yue TX. 2011. Surface modeling: high accuracy and high speed methods[M]. New York: CRC Press.

Zhao MW, Yue TX, Zhao N, et al. 2013. Spatial distribution of carbon stocks of forest vegetation in China based on HASM . Geographical Research, 68(9):1212–1224.

Zhao MW, Yue TX. 2016. Classification of high accuracy surface modeling (HASM) methods and their recent developments. Progress in Geography, 35(4):401–408.

Zhao N, Yue TX, Wang CL. 2015. Surface modeling of seasonal mean precipitation in China during 1951–2010. Progress in Geography, 70(3):447–460.

Zhao N, Yue TX, Zhao MW. 2014. Surface modeling of sunshine percentage in China based on a modified version of HASM. Geographical Research,33(7):1297–1305.

Zhu GF, He YQ, Pu T, et al. 2011. Spatial distribution and temporal Trends in potential evapotranspiration over Hengduan Mountains Region from 1960 to 2009. Acta Geographica Sinica, 66 (7):905–916.

Chapter 9
HASM-Based Downscaling Simulation of Temperature and Precipitation and Scenario Prediction in the Heihe River Basin

Based on spatial stationarity analysis, the downscaling simulation of the average annual temperature and precipitation in the Heihe River Basin is carried out by combining geographical factors, regression analysis and high-accuracy surface modelling (HASM). The differences between the downscaling results and the observation values at the stations are compared. The simulation accuracy of the proposed method and classical interpolation methods is also compared. The downscaling method is modified based on the downscaling method for historical period T1 (1976–2005) and the Coupled Model Intercomparison Project Phase 5 (CMIP5) outputs for the future periods T2 (2011–2040), T3 (2041–2070), and T4 (2071–2100) under Representative Concentration Pathway (RCP)2.6, RCP4.5, and RCP8.5 scenarios, and the downscaling equations of temperature and precipitation in future periods are obtained. Based on this, the downscaling simulation of the simulated annual average temperature and precipitation by CMIP5 model is carried out under the three scenarios. The results show that there is a good correlation between the results of the proposed downscaling method and the observation values, and its accuracy is higher than that of classical interpolation methods. The simulation results for future periods suggest that the fastest temperature increase is found under the RCP8.5 scenario. In the period of 2070–2100, the temperature can be greater than 10 °C in most areas except for the Qilian Mountains. Moreover, except for period from T3 to T4, the precipitation can decrease under the RCP2.6 scenario, while the precipitation shows an increasing trend in other periods under different scenarios. The largest increase in precipitation can occur from T1 to T2 under the RCP2.6 scenario, with an average of 28.07 mm.

N. Zhao and T. Yue, *High Accuracy Surface Modeling Method: The Robustness*, https://doi.org/10.1007/978-981-16-4027-8_9

9.1 Introduction

Climate change and human activities have led to temporal and spatial variability in hydrometeorological elements and have triggered changes in their statistical characteristics, which has had a series of impacts on hydrological and water resources and ecosystems (Mads et al. 2004; Zhang et al. 2010; Li et al. 2011). The study of the variation characteristics of temperature and precipitation is extremely important for an in-depth understanding of the water cycle (Li et al. 2013; Chen et al. 2014). Moreover, climate change is bound to affect future crop irrigation patterns to some extent because of the water balance relationship among precipitation, evaporation, and soil. The randomness, trends, periodicity and temporal rhythms of meteorological elements (e.g., precipitation, temperature, and radiation) can change in response to future climate change patterns, and the changes in these meteorological elements can directly affect the growth and development of crops (Li et al. 2010).

Located in northwestern China and originating from the northern slope of the Qilian Mountains, the Heihe River is the largest inland river in China and is the backbone of the oasis in the Hexi Corridor and the barrier against desert invasion in the north. In the context of global climate change, the climate characteristics of the Heihe River basin have changed significantly. The ecological environment, hydrological and water resources, crop water demands, and comprehensive management in the basin are closely related to climate change (Cao et al. 2003; Ning et al. 2008; Chen et al. 2014). Therefore, obtaining high-quality and high-precision spatial distribution data of temperature and precipitation in future scenarios is of great significance for the study of watershed ecosystems, agriculture, and water resources.

The temperature and precipitation data used in current ecological and hydrological studies are mainly meteorological observation data, satellite remote sensing data, and global model data, and the global model data are used for prediction in future scenarios. Currently, the simulation of meteorological elements in the Heihe River Basin in future scenarios is still lacking. Most of the existing studies focus on the climate change trends in the Heihe River Basin over the past decades or centuries(Li et al. 2009; Ding er al. 2009; Li et al. 2011; Sun et al. 2011). Although the general circulation models (GCMs) are able to provide macroscopic analysis of future temperature and precipitation changes on a global scale, the output spatial resolution is low, making the output unsuitable for climate response studies of various ecological and hydrological processes on a regional scale. In the prediction and early warning of water resources in river basins, particularly when distributed hydrological models are used, temperature and precipitation information with a high resolution are needed, yet the output of GCM does not meet this requirement. Statistical downscaling is an effective way to solve the conversion from low-resolution data to basin-scale data (Jeong et al. 2012). The basic principle is to use statistical empirical methods to establish linear or nonlinear relationships between large-scale meteorological variables and regional meteorological variables (Wilby et al. 2002). Nonetheless, the output accuracy and resolution of the existing methods are still unable to meet practical requirements, and an effective method is still lacking (Chen et al. 2011).

To address the above problems, this study constructs a scale conversion function using Coupled Model Intercomparison Project Phase 5 (CMIP5) data based on spatial stationarity and non-stationarity analyses and corrects the random errors introduced during downscaling by combining the meteorological observation data and using HASM (Yue et al. 2011,2013), a recently widely used method. A downscaling model for the average temperature and precipitation in the Heihe River Basin is established to realize the downscaling simulation of low-resolution temperature and precipitation data simulated by the Intergovernmental Panel on Climate Change (IPCC) CMIP5 model in future scenarios.

9.2 Data and Methods

9.2.1 Data

The data used to simulate the future climate scenario in the Heihe River Basin are the average results of 21 CMIP5 global climate models (Moss et al. 2008). The results are all downscaled to a spatial resolution of $1° \times 1°$ via interpolation (referred to as the CMIP5 dataset). The monthly average data are calculated using the simple average method, including the long-term historical precipitation and temperature data T1(1976–2005) and the precipitation and temperature of future periods (i.e., T2(2011–2040), T3(2041–2070) and T4(2071–2100)) under three different scenarios (i.e., RCP2.6, RCP4.5 and RCP 8.5). The CO_2 concentration is the highest in the RCP8.5 scenario and the lowest in the RCP2.6 scenario (Wang et al. 2012). The measurement data are obtained from 34 benchmark meteorological stations in and around the Heihe River Basin out of 752 stations nationwide (Fig. 9.1) and are processed to obtain the multi-year average values. The digital elevation model (DEM) data are from the United States Geological Survey (USGS) (http://srtm.csi.cgiar.org) and have a resolution of 90 m. The DEM data are resampled to a resolution of 1 km, and the spatial distribution of the DEM in the Heihe River Basin is extracted (Fig. 9.1).

9.2.2 Methods

From the CMIP5 dataset, the average temperature and precipitation raster data within a rectangle area containing the Heihe River Basin are extracted. The time periods for the data are 1976–2005 (T1), 2011–2040 (T2), 2041–2070 (T3), and 2071–2100 (T4). Geographically weighted regression (GWR) (Kamarianakis et al. 2008) with the same explanatory variables and sample points as ordinary least squares (OLS) regression is used to obtain the surfaces of the intercept, latitude regression coefficients, and elevation regression coefficients, and then the coefficient of variation

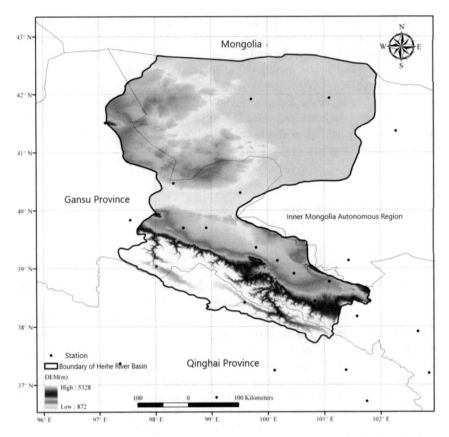

Fig. 9.1 Spatial distribution of the meteorological stations and DEM in the Heihe River Basin

(CV) is calculated as follows:

$$CV = \frac{\sigma}{\bar{Z}} \times 100\% \tag{9.1}$$

where σ is the standard deviation and $\bar{Z}$ is the average value of its attributes. When CV <10% , the variability is weak; when CV >100%, the variability is strong; and when 10% < CV < 100%, the variability is moderate. When there is strong variability, the spatial distribution of a variable is non-stationary, while in other cases, the spatial distribution of a variable is considered to be stationary (Lei 1988).

The results show that the CVs of the intercept, latitude regression coefficient and elevation regression coefficient are 24.66%, 23.36% and 21.30%, respectively. They are all under 25%, indicating a moderately weak variability. Thus, the spatial distribution of annual average temperature is relatively stable. Therefore, the background trend values of the temperature can be fitted using OLS regression.

The average historic temperature data in the CMIP5 dataset are superimposed with the DEM raster data with a spatial resolution of 1 km × 1 km. After projection transformation, the elevation of the model points is obtained, and the latitude and longitude coordinates (x, y) of the point are calculated. Regression analysis is carried out for the correlation between the average temperature and elevation and latitude. The following regression equation is obtained:

$$RMAT(x, y, T1) = 41.585 - 10^{-5} \cdot (0.982y - 404.090 Ele(x, y)) \qquad (9.2)$$

where $RMAT(x, y, T1)$ is the regression trend value of the original annual average temperature in the CMIP5 dataset in the T1 period, y represents the latitude (m), and $Ele(x, y)$ represents the elevation of the raster (x, y) (m).

Using the meteorological observation data, the corrected residuals of the regression trend values of the original annual average temperature in CMIP5 dataset are calculated, and the residuals at the meteorological stations are spatially interpolated using HASM. The residual surface with a spatial resolution of 1 km × 1 km is expressed as $\varepsilon = HASM(MAT_j - OMAT_j)$, and the downscaling equation for the annual average temperature in CMIP5 dataset can be expressed as

$$DMAT(x, y, T1) = RMAT(x, y, T1) + HASM(MAT_j - OMAT_j) \qquad (9.3)$$

where $DMAT(x, y, T1)$ is the downscaling result of historical annual average temperature, $OMAT_j$ represents the original annual average temperature in the CMIP5 dataset at meteorological station j, and MAT_j is the observed value of the annual average temperature at the meteorological station.

For the original annual average precipitation in CMIP5 dataset, surfaces of five regression coefficients (i.e., intercept, longitude, latitude, elevation, and aspect) are obtained via GWR. According to Eq. (9.1), the CVs of the regression surfaces for aspect and longitude are above 100%, indicating strong variability. The absolute values of the CVs of the other regression surfaces are greater than 40%, suggesting moderate variability. The relationships between precipitation and intercept, longitude, latitude, elevation, and aspect are non-stationary at 0.01 significance level. Thus, the relationships between precipitation in the Heihe River Basin and its explanatory factors are complex and spatially non-stationary and cannot be expressed by a single OLS global model. Therefore, the background trend of precipitation should be fitted using GWR.

To avoid the presence of extreme values during simulation, the model data are first normalized:

$$P(x, y, T1) = \frac{OMAP(x, y, T1)}{MAX\{OMAP(x, y, T1)\}} \qquad (9.4)$$

where $P(x, y, T1)$ is the normalized value of the observed annual average precipitation at (x, y) in the T1 period and $OMAP(x, y, T1)$ is the original annual average precipitation in the CMIP5 dataset in the T1 period.

To ensure data normality (Box 1964), the Box-Cox (BC) transformation is performed on $P(x, y, T1)$.

$$\Psi_{0.4}(P(x, y, T1)) = \left((P(x, y, T1))^{0.4} - 1\right)/0.4 \tag{9.5}$$

The original annual average precipitation in the CMIP5 dataset after BC transformation is superimposed with DEM and aspect raster data with a spatial resolution of 1 km × 1 km, and projection transformation is then carried out. GWR is used to fit the original annual average precipitation in the CMIP5 dataset, and the trend value is obtained:

$$\begin{aligned}\Psi_{0.4}(Ps(x, y, T1)) &= \theta_0(x, y) + \theta_1(x, y) \cdot x + \theta_2(x, y) \cdot y \\ &+ \theta_3(x, y) \cdot Ele(x, y) + \theta_4(x, y) \cdot ICA(x, y)\end{aligned} \tag{9.6}$$

where $x, y, Ele(x, y)$, and $ICA(x, y)$ are the longitude (m), latitude (m), elevation (m), and aspect, respectively, and $\theta_0(x, y), \theta_1(x, y), \theta_2(x, y), \theta_3(x, y)$, and $\theta_4(x, y)$ are constant terms and coefficients of the corresponding explanatory variables, which vary with geographic location.

The trend value is corrected using observation data at meteorological stations, and the 1 km × 1 km residual surface optimized by HASM and the downscaled BC transformation of annual average precipitation in the CMIP5 dataset are obtained.

$$\begin{aligned}\varepsilon(x, y, T1) &= HASM(\Psi_{0.4}(MAP_j/MAX\{OMAP_j(x, y, T1)\}) \\ &- \Psi_{0.4}(Ps(x, y, T1))\end{aligned} \tag{9.7}$$

$$\Psi_{0.4}(DMAP(x, y, T1)) = \Psi_{0.4}(Ps(x, y, T1)) + \varepsilon(x, y, T1) \tag{9.8}$$

where MAP_j is the annual average precipitation observed at meteorological station j, $OMAP_j$ represents the annual average precipitation in the CMIP5 dataset at meteorological station j, and $DMAP(x, y, T1)$ represents the downscaling result of the annual average precipitation in the T1 period.

The inverse BC transformation and inverse normalization are performed to obtain the downscaled mapping of the annual average precipitation in the CMIP5 dataset.

$$DMAP(x, y, T1) = MAX\{OMAP(x, y, T1)\} \cdot (0.4 \cdot \Psi_{0.4}(P(x, y, T1)) + 1)^{\frac{1}{0.4}} \tag{9.9}$$

9.3 Comparison of the CMIP5 Baseline Data and Meteorological Observation Data

To compare the CMIP5 simulation results with the meteorological observation data and verify the improvement in the downscaling process on the CMIP5 simulation results, the $1° \times 1°$ annual average temperature and precipitation data in CMIP5 dataset from 1976 to 2005 and the 1 km $\times$ 1 km annual average temperature and precipitation data after the downscaling process are compared with the observation data at meteorological stations using cross validation. The mean absolute error (MAE), mean relative error (MRE), and root mean square error (RMSE) of the CMIP5 data and downscaling data at each station are shown in Table 9.1.

$$RMSE = \sqrt{\frac{\sum_{k=1}^{N}(f_k - \overline{f}_k)^2}{N}}, MAE = \sum_{k=1,\cdots,N} \frac{|f_k - \overline{f}_k|}{N},$$

$$MRE = \sum_{k=1,\cdots,N} \frac{|f_k - \overline{f}_k|}{f_k N}$$

where f_k is the CMIP5 mode value or DMAT (DMAP), $\overline{f}_k$ represents the observation value OMAT (OMAP) at the meteorological station, and N is the number of stations.

The CMIP5 model exhibits poor accuracy in simulating the annual average temperature and annual average precipitation. After downscaling, the accuracy is greatly improved.

Figure 9.2 shows a comparison between the observed annual average temperatures and the simulated annual average temperatures by CMIP5 and the CMIP5 outputs after downscaling. The results show that the observed average temperature in the Heihe River Basin from 1976 to 2005 is 5.65 °C, and the CMIP5 value is 1.56 °C; CMIP5 underestimates the temperature in the Heihe River Basin by an average of 4.09 °C. After downscaling, the simulation values are closer to the actual observed values.

Figure 9.3 shows a comparison between the observed annual average precipitation and the simulated annual average precipitation by CMIP5 and the CMIP5 outputs after downscaling. The results show that the observed average precipitation in the Heihe River Basin is 158.22 mm and that the average value simulated by CMIP5

Table 9.1 Comparison of the CMIP5 output and downscaling results

Method	Temperature			Precipitation		
	MAE (°C)	MRE (%)	RMSE (°C)	MAE (mm)	MRE (%)	RMSE (mm)
CMIP5	4.4	74.48	2.93	276.99	326.7	291.76
Downscaling	0.13	3.00	0.19	5.75	8.86	73.93

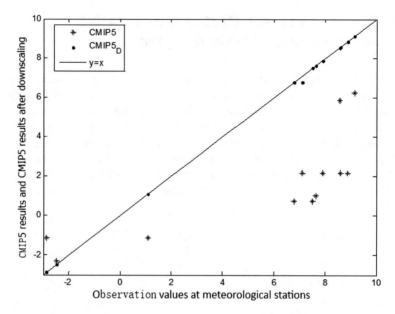

Fig. 9.2 Comparison between the average temperatures of the observation values and CMIP5 results and the CMIP5 results after downscaling

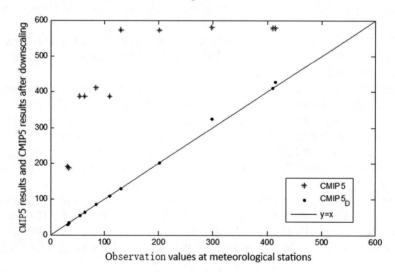

Fig. 9.3 Comparison between the average precipitation observation values and CMIP5 results and the CMIP5 results after downscaling

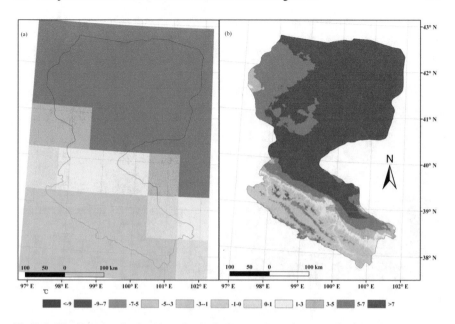

Fig. 9.4 Distributions of temperature for the CMIP5 results and downscaled CMIP5 results

is 435.21 mm. CMIP5 overestimates the precipitation in the Heihe River Basin by an average of 276.99 mm. After downscaling, the results are closer to the actual observed values, and the accuracy after downscaling is significantly improved.

Figure 9.4 shows the CMIP5 results and downscaling results for the annual average temperature in the Heihe River Basin from 1976 to 2005. The CMIP5 data cannot characterize the spatial distribution of temperature in the basin due to the low resolution. In contrast, the results after downscaling can reflect the spatial distribution of average temperature well due to the introduction of geographical factors, observation values, and other information.

Figure 9.5 shows the CMIP5 results and downscaled CMIP5 results for annual average precipitation from 1976 to 2005. The simulation results of CMIP5 can hardly be used in the study of eco-hydrological process in the Heihe River Basin. After downscaling, the detailed information of the average precipitation becomes obvious, and the downscaled CMIP5 results can better reflect the spatial distribution characteristics of the annual average precipitation.

To investigate the accuracy of the downscaling method in this study, the temperature and precipitation are simulated using eleven methods (i.e., the inverse distance weighting (IDW) method, kriging, spline, OLS-IDW, GWR-BC-IDW, OLS-kriging, GWR-BC-kriging, OLS-spline, GWR-BC-spline, OLS-HASM, and GWR-BC-HASM). Tables 9.2 and 9.3 are the MAE and MRE of different methods.

The classical interpolation methods have large simulation errors. After the introduction of GWR and residual correction, the simulation accuracy is significantly improved. For the multi-year average temperature, the OLS-HASM method is more accurate than the other methods, followed by the OLS-spline method. For the annual

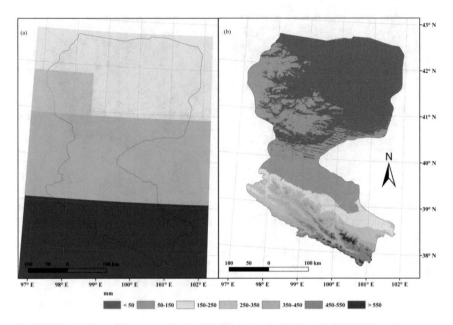

Fig. 9.5 Distributions of precipitation for CMIP5 results and downscaled CMIP5 results

Table 9.2 Comparison of different downscaling methods for CMIP5 temperature data

Error	IDW	Kriging	Spline	OLS-IDW	OLS-kriging	OLS-spline	OLS-HASM
MAE (°C)	17.13	17.38	17.27	0.93	0.28	0.18	0.13
MRE (%)	241.83	245.34	243.88	13.15	3.95	3.49	3.00

Table 9.3 Comparison of different downscaling methods for CMIP5 precipitation data

Error	IDW	Kriging	Spline	GWR-IDW	GWR-kriging	GWR-spline	GWR-HASM
MAE (mm)	221.48	239.01	232.05	13.73	5.85	432.19	5.75
MRE (%)	347.96	373.90	363.02	21.48	9.15	224.09	8.86

average precipitation, the accuracy of GWR-HASM method is the highest, and GWR-kriging method is second. The high accuracy of HASM is due to an initial driving field in the iterative process. To obtain accurate results, the spline interpolation results are used as the HASM driving field in the downscaling simulation of multi-year average temperature, and kriging results are used as the HASM driving field in the simulation of average precipitation. An obvious advantage of HASM is that it integrates the results of other interpolation methods and combines them with actual observation data to achieve a high accuracy.

9.4 Downscaling Simulation of Future Temperature and Precipitation in the Heihe River Basin Under Different Scenarios

The above results suggest that the downscaling method proposed in this study is effective; compared with the actual observation values at the stations, the error is small, and the accuracy is better than those of other methods. Therefore, based on the downscaled CMIP5 data of 1976–2005, the annual average temperature and precipitation for the periods of 2011–2040, 2041–2070, and 2071–2100 are calculated. After the CMIP5 model baseline data is added to the changes under the RCP scenarios for the periods of 2011–2040, 2041–2070, and 2071–2100, the corrected downscaled results of future annual average temperature and annual average precipitation are obtained. The correction equation is expressed as follows:

$$SMAT(x, y, t_k) = DMAT(x, y, T1) \\ + HASM(OMAT(x, y, T_k) - OMAT(x, y, T1)) \quad (9.10)$$

$$SMAP(x, y, T_k) = DMAP(x, y, T1) \\ + HASM(OMAP(x, y, T_k) - OMAP(x, y, T1)) \quad (9.11)$$

where T_k for which $k = 2$, 3, and 4, represents the periods of 2011–2040, 2041–2070 and 2071–2100, respectively. $SMAT(x, y, T_k)$ and $SMAP(x, y, T_k)$ represent the corrected annual average temperature and annual average precipitation in the $T_k(k = 2$, 3, and 4) periods. $DMAT(x, y, T1)$ and $DMAP(x, y, T1)$ represent the downscaled CMIP5 results of annual average temperature and annual average precipitation in 1976–2005, respectively. $OMAT(x, y, T_k)$ and $OMAP(x, y, T_k)$ are the original CMIP5 results of annual average temperature and annual average precipitation, respectively.

For the annual average temperature and annual average precipitation data in the Heihe River Basin during T2 (2011–2040), T3 (2041–2070), and T4 (2071–2100) under the three scenarios (RCP2.6, RCP4.5, and RCP8.5), correction and high-accuracy downscaling are performed to obtain the annual average temperature and precipitation data under RCP2.6, RCP4.5, and RCP8.5 in a 1-km grid. Using ArcGIS, the temporal and spatial distributions of the average temperature and precipitation are obtained.

Figures 9.6, 9.7 and 9.8 are the downscaled simulation results of the annual average temperatures during three periods (i.e., 2011–2040, 2041–2070, and 2071–2100) under scenarios RCP2.6, RCP4.5, and RCP8.5. The temperature increases the fastest under the RCP8.5 scenario. The temperature is greater than 10 °C in most areas except the Qilian Mountains during 2070–2100. The ArcGIS analysis shows that under the RCP2.6 scenario, from T1 to T2, the annual average temperature in the Heihe River Basin increases by 1.80 °C, and the largest temperature increase (2.02 °C) occurs in the area near the Binggou Watershed of the Qilian Mountains; from T2 to T3, the

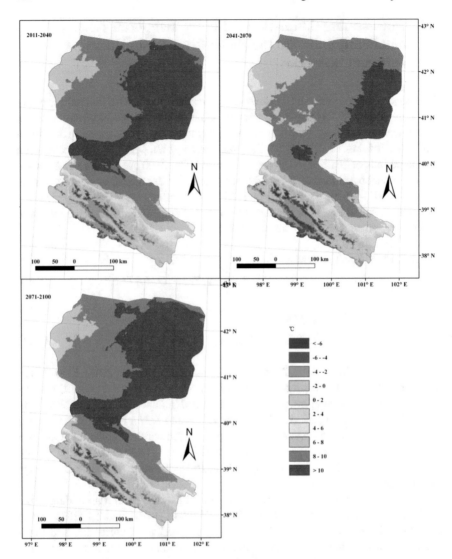

Fig. 9.6 Downscaled results of temperature in different periods under the RCP2.6 scenario

annual average temperature in the Heihe River Basin shows a decreasing trend, with an average decrease of 0.67 °C, and the largest temperature decrease occurs in the southern and western areas of the Heihe River Basin, with a maximum drop of 0.9 °C; and from T3 to T4, the annual average temperature in the Heihe River Basin increases by 0.83 °C, and the area with largest temperature increase (1.24 °C) is in the eastern part of the lower reaches. Under the RCP4.5 scenario, from T1 to T2, the annual average temperature in the Heihe River Basin increases by an average of 2.45 °C, and the largest temperature increase (2.80 °C) occurs in the eastern part of

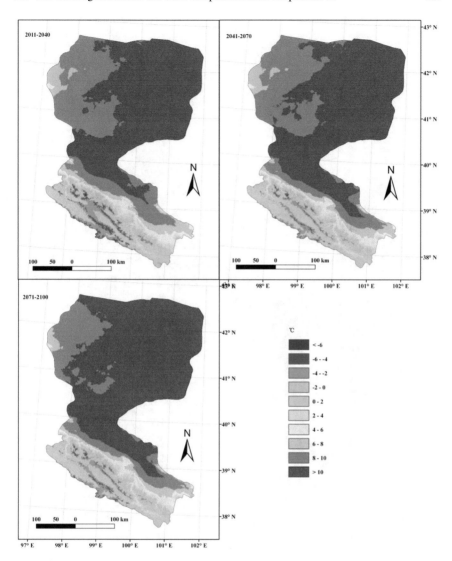

Fig. 9.7 Downscaled results of temperature in different periods under the RCP4.5 scenario

the lower reaches; from T2 to T3, the average temperature in the Heihe River Basin increases by 0.23 °C, with a maximum temperature increase of 1.01 °C; and from T3 to T4, the annual average temperature in the Heihe River Basin increases by 0.41°C on average, and the largest temperature increase (1.02 °C) occurs in the northern part of the lower reaches of the basin. Under the RCP8.5 scenario, from T1 to T2, the annual average temperature in the Heihe River Basin increases by 1.47 °C on average, and the largest increase (1.81 °C) occurs in the southwest of the Qilian Mountains; from T2 to T3, the annual average temperature in the Heihe River Basin increases by

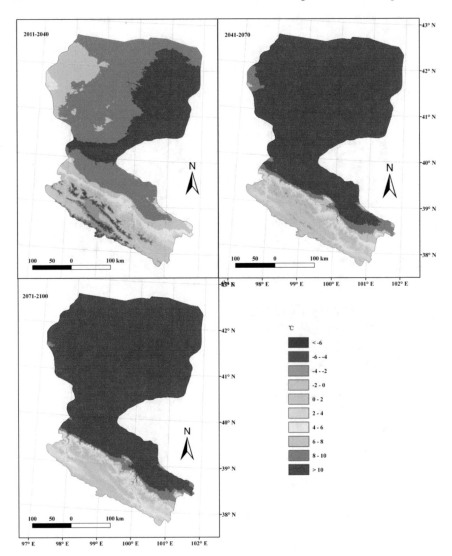

Fig. 9.8 Downscaled results of temperature in different periods under the RCP8.5 scenario

3.32 °C on average, and the maximum temperature increase (3.79 °C) occurs in most parts of the Qilian Mountains; and from T3 to T4, the basin temperature increases by 1.55 °C on average, with a maximum value of 1.89 °C, and the area with the largest temperature increase is the northeastern part of the lower basin.

Figures 9.9, 9.10, and 9.11 are the GWR + HASM downscaled results of annual average precipitation during three periods (i.e., 2011–2040, 2041–2070, and 2071–2100) under scenarios RCP2.6, RCP4.5, and RCP8.5.

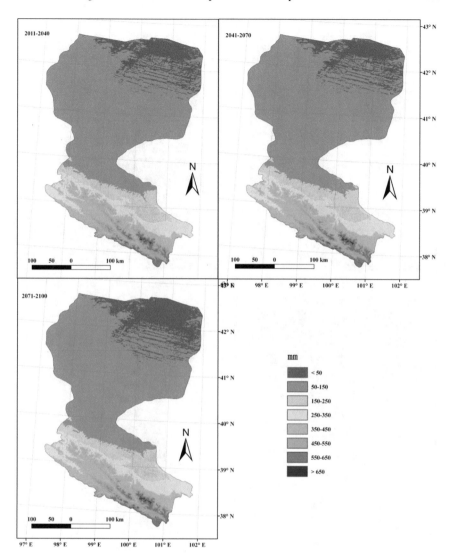

Fig. 9.9 Downscaled results of precipitation in different periods under the RCP2.6 scenario

Under the RCP2.6 scenario, from T1 to T2, the average increase in precipitation in the Heihe River Basin is 28.07 mm, and the largest increase (47.39 mm) occurs in the southeastern Qilian Mountains; from T2 to T3, the average increase in precipitation in the Heihe River Basin is 2.50 mm, with a maximum increase of 14.37 mm in the southeastern Qilian Mountains; and from T3 to T4, the average precipitation decreases by 1.93 mm, and the largest decrease (4.73 mm) occurs in the southeastern Qilian Mountains. Under the RCP4.5 scenario, from T1 to T2, the average increase in precipitation in the Heihe River Basin is 17.30 mm, and a maximum value of

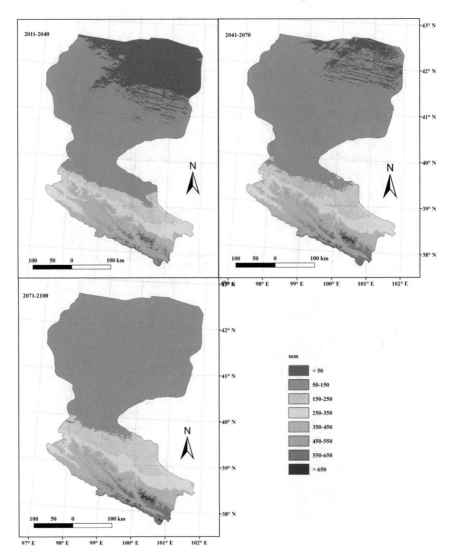

Fig. 9.10 Downscaled results of precipitation in different periods under the RCP4.5 scenario

32.36 mm occurs in the eastern Qilian Mountains; from T2 to T3, the average increase
in precipitation is 17.68 mm, and the area with the largest increase (35.68 mm)
is the southeastern Qilian Mountains; and from T3 to T4, the average increase in
precipitation is 12.50 mm, and the area with the largest increase (23.34 mm) is
the northwestern Qilian Mountains. Under the RCP8.5 scenario, from T1 to T2,
the average increase in precipitation in the Heihe River Basin is 25.08 mm, with
a maximum increase of 46.14 mm in the Qilian Mountains; from T2 to T3, the
average increase in precipitation in the Heihe River Basin is 23.47 mm, and the

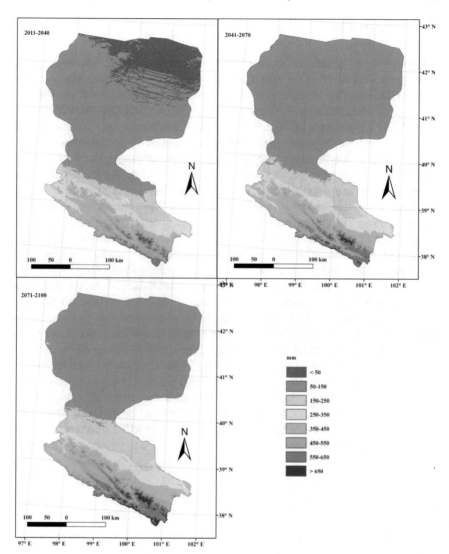

Fig. 9.11 Downscaled results of precipitation in different periods under the RCP8.5 scenario

largest increase (43.54 mm) occurs in the eastern Qilian Mountains; and from T3 to T4, the average increase in precipitation is 23.13 mm, with a maximum increase of 50.40 mm in the western Qilian Mountains.

9.5   Conclusion

Based on spatial stationarity analysis, this study proposes a statistical downscaling equation for the CMIP5 model data to address the low-resolution problem. Based on the proposed equation, downscaling of the simulated temperature and precipitation in the Heihe River Basin in the future periods of 2011–2040 (T2), 2041–2070 (T3), 2071–2100 (T4) under the RCP2.6, RCP4.5, and RCP8.5 scenarios is carried out. The differences between the proposed method and the measured values at meteorological stations and the simulation accuracy differences between the proposed method and the classical interpolation methods are compared. To study the hydrological and water resources and ecosystem in the Heihe River Basin, this study achieves the prediction of meteorological elements in the Heihe River Basin in future scenarios by downscaling. The results show that the multi-year annual average temperature can be simulated by using OSL regression and HASM residual correction to downscale the CMIP5 data, while the average precipitation data can be simulated by using GWR and HASM residual interpolation for downscaling the CMIP5 output. The downscaling results of the proposed method are close to the actual observation values at the meteorological stations, and the accuracy is higher than those of classical interpolation methods. The simulation results of temperature and precipitation in future periods show that the annual average temperature in the Heihe River Basin is increasing under different scenarios, except for a decrease from T2 to T3 under the RCP2.6 scenario. Under the RCP8.5 scenario, the temperature rises the fastest, and during the period of 2070–2100, the temperature in most areas except the Qilian Mountains is above 10 °C. Except for the T3-T4 period under the RCP2.6 scenario, the precipitation tends to increase in all periods under different scenarios. The largest increase in precipitation is in the T1-T2 period under the RCP2.6 scenario, with an average increase of 28.07 mm.

References

Box GEP, Cox DR. 1964. An analysis of transformation. Journal of the Royal Statistical Society, 26: 211–252.

Cao L, Dou YX, Zhang DY. 2003. Impact of climate change on ecological environment in Heihe River Basin. Drought weather, 21(4): 45–49.

Chen ST, Tseng HW, Lin CY, et al. 2011. Hydrological drought in Tseng-Wen reservoir basin under climate change scenarios. Journal of Taiwan Agricultural Engineering, 57 (3): 44–60.

Chen YN, Li Z, Fan YT, et al. 2014. Research progress on the impact of climate change on water resources in the arid region of Northwest China. Acta Geographica Sinica, 69(9): 1295–1304.

Ding R, Wang FC, Wang J, er al. 2009. Analysis on Spatial-temporal Characteristics of Precipitation in Heihe River Basin and Forecast Evaluation in Recent 47 Years. Journal of Desert Research, 29(2): 335–341.

Jeong DL, St-Hilaire A, Ouarda TBMJ, et al. 2012. Comparison of transfer functions in statistical downscaling models for daily temperature and precipitation over Canada. Stochastic Environmental Research and Risk Assessment, 26 (5): 633–653.

Kamarianakis Y, Feidas H, Kokolatos G, et al. 2008. Evaluating remotely sensed rainfall estimates using nonlinear mixed models and geographically weighted regression. Environment Modeling & Software, 23:1438–1447.

Lei ZD. 1988. Soil dynamics. Tsinghua University Press.

Li FP, Zhang GX, Dong LQ. 2013. Studies for Impact of Climate Change on Hydrology and Water Resources. Scientia Geographica Sinica, 2013, 33(4):457–464.

Li HY, Wang KL, Jiang Y, et al. 2009. Research progress and Prospect of precipitation in Heihe River Basin. Journal of Glaciology and Geocryology, 31(2): 334–341.

Li XD, Fu H, li FX, et al. 2011. Research progress on effects of climate change on ecological environment in Northwest China. Pratacultural Science, 2011, 2: 286–295.

Li Y, Yang XG, Wang WF, et al. 2010. The Possible Effects of Global Warming on Cropping Systems in China V. The Possible Effects of Climate Warming on Geographical Shift in Safe Northern Limit of Tropical Crops and the Risk Analysis of Cold Damage in China. Scientia Agricultura Sinica, 43(12):2477–2484.

Li ZL, Xu ZX. 2011. Analysis of the abrupt change of temperature and precipitation in Heihe River Basin in the past 50 years. Resource Science, 33(10): 1877–1882.

Mads CF, Eric P. 2004. Using large-scale climate indices in climate change ecology studies. Population Ecology, 46(1): 1–12.

Moss R, Babiker M, Brinkman S, et al. 2008. Towards new scenarios for analysis of emissions, climate change, impacts, and response strategies. Technical Summary, Intergovernmental Panel on Climate Change, Geneva.

Ning BY, He YQ, He XZ, et al. 2008.Advances on Water Resources Research in Heihe River Basin. Journal of Desert Research, 28(6): 1180–1185.

Sun J, Jiang Y, Wang KL, et al. 2011.The Fine Spatial distribution of mean precipitation and the estimation of total precipitation in Heihe River Basin . Journal of Glaciology and Geocryology, 33(2): 318–324.

Wang SW, Luo Y, Zhao ZC, et al. 2012. A new generation of greenhouse gas emission scenarios. Progressus Inquisitions De Mutatione Climates, 8(4): 305–307.

Wilby RL, Dawson CW, Barrow EM. 2002. SDSM- a decision support tool for the assessment of regional climate change impacts. Environmental Modeling & Software, 17(2): 147–159.

Yue TX, Zhao N, Ramsey RD, et al. 2013. Climate change trend in China, with improved accuracy. Climatic Change, 120:137–151.

Yue TX. 2011. Surface Modeling: High Accuracy and High Speed Methods. New York: CRC Press.

Zhang YL, Ouyang H, Zhang XZ, et al. 2010. Vegetation change and its responses to climatic variation based on eco-geographical regions of Tibetan Plateau.Geographical Science, 29(11): 2004–2016.

Printed in the United States
by Baker & Taylor Publisher Services